Surface Tension Supplement №2

WHAT REMAINS OF A BUILDING DIVIDED INTO EQUAL PARTS AND DISTRIBUTED FOR RECONFIGURATION

Surface Tension Supplement № 2

EDITED BY KEN EHRLICH & BRANDON LABELLE

ERRANT BODIES PRESS 2009

**WHAT REMAINS OF A BUILDING DIVIDED
INTO EQUAL PARTS AND DISTRIBUTED FOR RECONFIGURATION**

Surface Tension Supplement № 2

Edited by Ken Ehrlich & Brandon LaBelle

ISBN: 978-0-9772594-3-4

Errant Bodies Press 2009

Design: James W Moore & Penny Pehl / tenderling.com

ERRANT BODIES
Lettestrasse 7   1835 Wollam Street
10437 Berlin   Los Angeles, CA 90065
Germany   USA

www.errantbodies.org

Errant Bodies publications are distributed by DAP, New York
www.artbook.com

Ken Ehrlich & Brandon LaBelle

Introduction

The second publication in the *Surface Tension Supplement* series marks the ongoing evolution of a platform for publishing and organizing exhibitions related to international contemporary spatial practices. Initially developed as a comprehensive anthology devoted to documenting a broad range of site-specific work in art, architecture and performance and reflecting critically on the history of the notion of "site", the project has blossomed into a vehicle for engaging modes of locational exchange across a range of disciplines and media.

The current volume in the series takes as its title a modified and slightly subverted version of another title—the title of an essay by Jean Genet, *What Remains of a Rembrandt Torn Into Four Equal Pieces and Flushed Down the Toilet*. In this text, Genet muses on the complexity of emotion found in a momentary, shared glance on a train and what can happen in the eyes and body of a viewer gazing at late Rembrandt paintings. He states, "A work of art should exalt only those truths which are not demonstrable, and which are even 'false,' those which we cannot carry to their ultimate conclusions without absurdity, without negating both them and ourself." Considering the subtle eroticism of the Rembrandt paintings, Genet reminds the reader of the complex ways that subjective experience and social formations mingle and co-contaminate. By trading the *Rembrandt* of the title for *buildings* and *flushing* for *reconfiguration*, we hope to draw attention to the built environment as both a dominant force in shaping spatial experience and yet one that is malleable and changing. At the heart of our on-going project related to "site-based" practices is an interest in exploring instances of intersections of the individual with the social and the private with the public. Might Genet's formulations of an artistic experience that exalts certain truths even to the point of negation inform a perspective on spatial thinking that incorporates the very subversion of its own assumptions? A view which encourages unorthodox strategies and articulations, practices that, in seeking to occupy or activate the site of their own production or performance, might come to negate their own "truth"? We resist the temptation to answer such questions prescriptively and instead hope to multiply and complicate the ongoing conversations about spatiality.

This volume is primarily focused on questions of built space with a view toward engaging its social and political dimensions. Ava Bromberg presents a personal account of her work in an urban planning office in the Chinese city of Shenzhen and reflects on the scale, pace and challenges of contemporary urban planning in China in general. Issues pertaining to contemporary urbanism and the related excesses of current building are brought forward, underscoring the ongoing tension between plan and use, concept and environment. A broad range of spatial questions are further explored by Jesko Fezer and Mathias Heyden in their analysis of participatory and interventionist architecture in Europe from 1945 to the present. This historical analysis presents a lineage of idiosyncratic spaces and plans that necessarily incorporate informal and often surprisingly generative methods of producing place. Carl Michael von Hausswolff, a Swedish artist working with the built environment, details the process of designing and building a house on The Land Foundation site in Chiangmai, Thailand. Through his reflections, an intimate portrait of The Land Foundation project emerges and readers are introduced to a community that takes a holistic approach toward social existence. Hausswolff's house project becomes an embodiment of the community's spirit of generosity, in which housing is both a vehicle for the imagination as well as a form of gift giving, where anyone may use the house and make creative additions to the site in general. Extending notions of participation and urban planning, artist and activist Nis Rømer presents documentation of a series of gardening interventions made in Copenhagen in 2006, raising issues of urban farming, green spaces, and city politics. His *Hot Summer of Urban Farming* group project offered tangible proposals from different artists, uncovering the degree to which urban space is often perceived as immutable while productively engaging with the material processes of spatial exchange. Finally, Los Angeles-based architect Rachel Allen presents documentation of a recent project and dialogues with Ken Ehrlich on aspects of contemporary architecture and the convergences and divergences between the practices of art and architecture.

As the primary site or terrain where bodies interact with physical environments and abstract notions of space conflict with lived experience, the built environment is constantly coded and decoded through interaction and exchange. Throughout this publication, architecture is theorized as a practice that can bring social histories and personal experiences to bear on urban policy and planning, challenging the drive towards an often short-sighted professionalism. Although the language, history and discourses of art and architecture are distinct, and such differences always influence the cultural reception of each, the disciplinary boundaries between certain forms of artistic practice and critical work being done in the context of architecture share an overarching concern with what might be called the intersection of social and spatial configurations. The work in this *Supplement* explores the uses and misuses of contemporary architectural and urban space, and in doing so highlights the generative unpredictability built into spatial experience. Our aim is to mark these experiments with potential.

Biographies

RACHEL ALLEN grew up in San Francisco and received her architecture degrees from Princeton University. She went on to work with Gehry Partners, ultimately as Assistant Project Designer of the Stata Center at MIT. In 2002-2003 she received the Rome Prize in architecture, a yearlong fellowship. She currently serves on the Board of the Outpost for Contemporary Art and has taught at the Mountain School of Art, Princeton University, SCI-Arc, UCLA, UC Riverside, USC and Woodbury University. She is owner and principal of Rachel Allen Architecture, whose built work includes residential, retail, commercial and food service projects in Northern and Southern California, and collaborations with artists on large-scale installations.

AVA BROMBERG is an internationally recognized urban planning scholar, artist, and spacemaker. Her research and practice focus on creative economies, public and social space, community-driven planning, and responsible development. As a 2002–2003 Thomas J. Watson Fellow, Ava researched local incubators of culture, creativity, and community as public art in twenty cities around the world. She is a co-founder of Mess Hall, a storefront cultural center in Chicago, co-organizer of the Just Space(s) exhibition and symposium series, and co-editor of *Belltown Paradise/Making their own Plans*. Ava holds a Master's Degree in Urban Planning from UCLA, where she is a doctoral student.

KEN EHRLICH is an artist and writer based in Los Angeles. He has exhibited internationally in a variety of media including video, sculpture and photography. His work interweaves architectural, technological and social themes to play with ideas of invention and circumvention; superstructure and infrastructure; consumption and waste; and site, place and location. He often collaborates with architects and other artists in site-specific and/or community-based projects. He is the co-editor of *Surface Tension: Problematics of Site* (2003) and *Surface Tension Supplement No. 1* (2006). He currently teaches at The California Institute of the Arts (CalArts) and in the Department of Art at UC Riverside.

JESKO FEZER is co-manager of the specialist bookstore *Pro qm* in Berlin, Germany and co-editor of the political architecture magazine *An Architektur*. He was recently a Visiting Professor for Urban Research in the master's degree course at the Academy of Fine Arts in Nuremberg. As an architect *(ifau und Jesko Fezer)*, he realized re-designs and buildings for cultural institutions in Berlin, Munich, Stuttgart, Utrecht and Graz. His most recent publications are *Fezer/a42.org: Planungsmethodik gestern, Fezer/Heyden: Hier entsteht. Strategien partizipativer Architektur und räumlicher Aneignung, Fezer/ Schmitz: Lucius Burckhardt: Wer plant die Planung?*, and *Fezer/Reichard/Wieder: Martin Pawley's Garbage Housing*.

MATHIAS HEYDEN is a carpenter and an architect with *ISPARA* (Institute for Strategies of Participatory Architecture and Spatial Appropriation). He is the co-founder of K 77, a project dedicated to self-determined and cooperative dwelling, work and culture in a formerly squatted house. His work is based on a commitment to direct-democracy and solidarity-oriented, sustainable and holistic production of spaces and use of the built environment. He organized an exhibition on participatory architecture and edited a companion publication with Jesko Fezer entitled *Hier entsteht. Strategien partizipativer Architektur und räumlicher Aneignung* in 2004. Other publications include *Wohnen in eigener Regie! Gemeinschaftsorientierte Strategien für die Mieterstadt* (with Bildungswerk Berlin der Heinrich-Böll-Stiftung) and *An Architektur 19 – 21: Community Design. Involvement and Architecture in the US since 1963* (with *An Ar-chitektur*).

BRANDON LABELLE is an artist and writer working with sound and cultural narratives. He is the co-editor of the *Surface Tension* book series, and he manages Errant Bodies Press. His Prototypes for the Mobilization and Broadcast of Fugitive Sound was exhibited at Enrico Fornello gallery, Prato, in 2007, and at Ybakatu Gallery, Curitiba, Brazil in 2009. His ongoing project to build a library of radio memories was presented at Casa Vecina, Mexico City, in 2008. He is the author of *Background Noise: Perspectives on Sound Art* (Continuum 2006).

AROON PURITAT was born in Chiang Rai, Thailand and attended Silpakorn University in Bangkok, obtaining a B.Arch. in 1997. Since then he has realized several designs, including the "Annex House" in Chiang Rai (2001) and the "Mud House" at the Chiang Mai Art Museum (2002). Puritat's work involves the merging of art and architectural ideas. He also works as a freelance writer for the magazine Art4D and Wallpaper Magazine (Thai edition). He has recently completed Rirkrit Tiravanija's private house in Chiang Mai, Thailand and is currently working on a new book "THAI" as well as "MaMa house", a new studio for him and his mother in Chiang Mai.

NIS RØMER makes public art in the city, on the web and in the news media. He has a special interest in the social and political organization of space and in how processes of globalization affects the city and our natural environment. He often works in groups and is a co-founder of www.free-soil.org working with issues of sustainability, and of www.publik.dk making art in public spaces in Copenhagen.

CARL MICHAEL VON HAUSSWOLFF is an artist based in Stockholm. Since the end of the 1970s he has worked as a composer using the tape recorder as his main instrument and as a conceptual visual artist working with performance art, light- and sound installations and photography. His interest in architecture and topography has also resulted in the films "Hashima, Japan 2002" and "Al Qasr, Bahriyah Oasis, Egypt, 2005" with Thomas Nordanstad. In 1993 he and Leif Elggren developed the ever-lasting conceptual piece "The Kingdoms of Elgaland-Vargaland" and in 2005 he completed his Starhouse project for The Land Foundation, Chiangmai, Thailand.

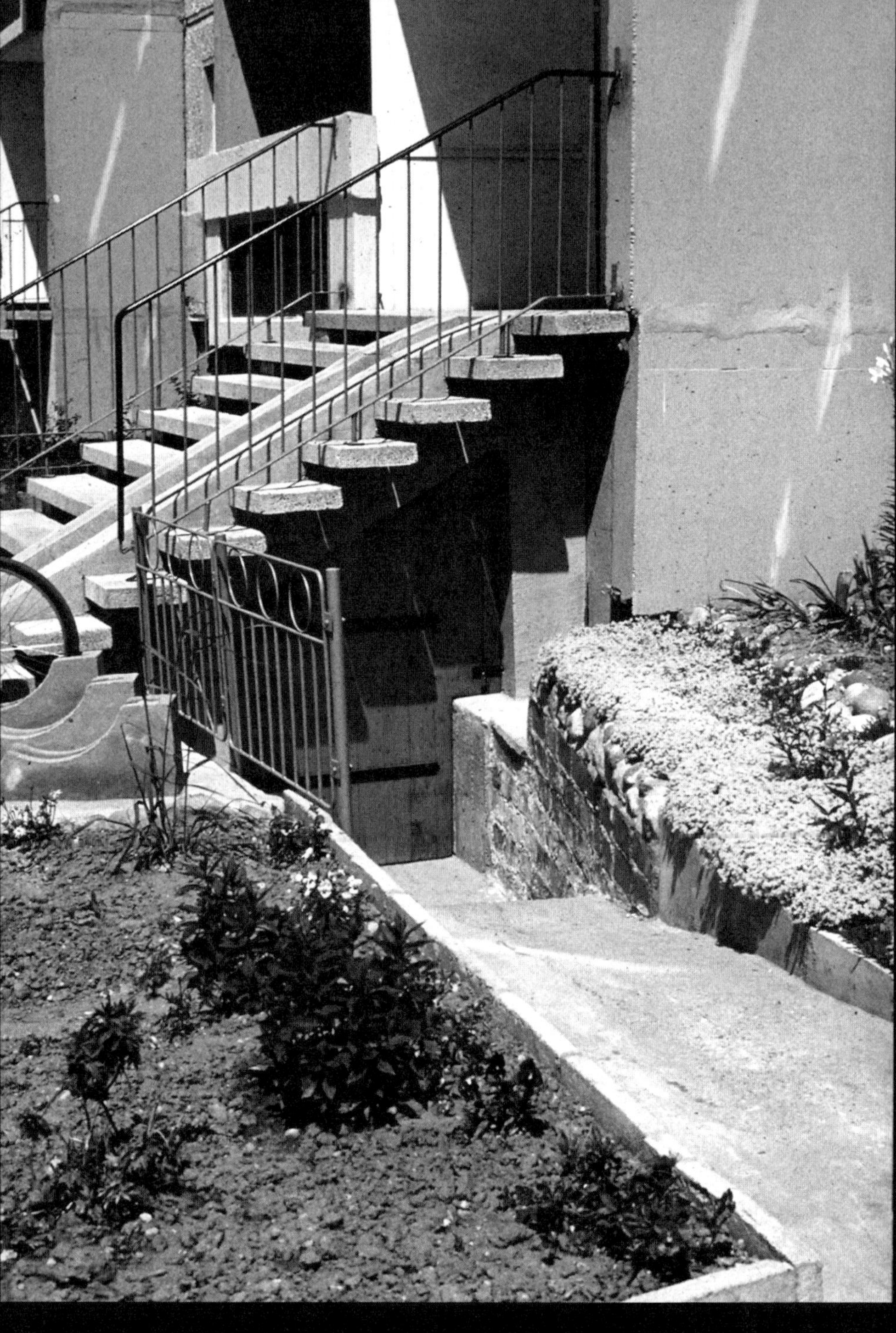

Jesko Fezer and Mathias Heyden

Under Construction
Strategies of Participative Architecture and Spatial Appropriation

Translation by Elisabeth Felicella and Ines Schaber

Clarence Schmidt built his house himself in the hill country near Woodstock, a simple wood cabin. In 1948 the learned mason began extending it piece by piece. Schmidt used only found materials—old windows, doors, wooden slats, planks, tarpaper—and over the years it grew into a labyrinthine, seven-story pyramid of thirty-five rooms. His intensely painted and highly decorative "House of Mirrors" was covered throughout with reflective materials like aluminium foil, mirrors and Christmas tree decorations. In 1968 it was completely destroyed by a fire, but through exhibitions and publications, Schmidt's building and his other forms of spatial appropriation—whether born of necessity or diversion—gained the notice of professional planners.

In 1946, 40,000 families occupied thousands of evacuated military camps in Britain, soon after health authorities legalised the settlements. In 1966 the sociologist Phillipe Boudon published an investigation of the Pessac housing project—entirely altered by its inhabitants in color, form and organization—which Le Corbusier had planned in 1926 for a French manufacturer. Since the 1960s, the architect John F.C. Turner has surveyed illegally established settlements in Peru and describes the superiority of individual and local resources to a bureaucratic centralized housing system.

Such considerations of unplanned production of space played an important role in international debate of the post-war period surrounding the renewal or dissolution of modernism. An increasing uncertainty about the function of design and a dissatisfaction with the paradigms of an industry-oriented, post-war modernism, spurred architects to search for new concepts. They set out on different paths toward the users

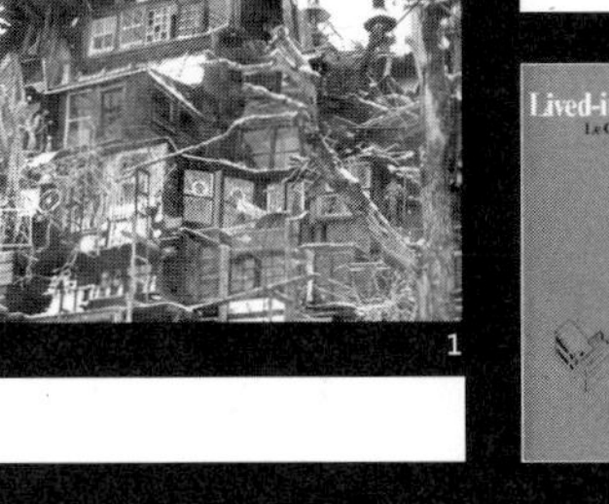

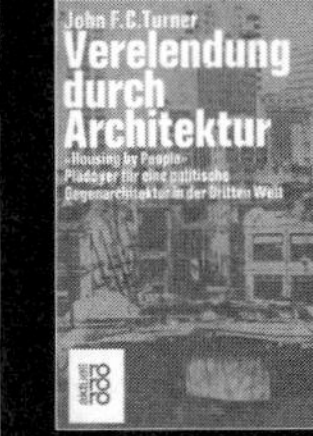

1 Schmidt: *House of Mirrors*, USA 1948–1968

2 & 3 Boudon: *Lived-in Architecture. Le Corbusier's Pessac revisited*, London 1972

4 Turner: *Housing by People*, London 1976

of their products. They encouraged, motivated, organised, sanctioned and outlined the participation of those affected by building: the inhabitants.

In this regard an architectural history of participation can be understood as a reaction to the real pressure of spatial acquisition. Participation is tested as an approach to forging new connections to a vital world and its everyday occurrences. Techniques of participation are always bound to social power relations. Models for incorporating inhabitants into architectural processes emerged just as the Fordist regulation of labor and lifestyle reached its limits and a new flexible regime of work and consumption emerged, integrating the employee into a more dynamic hierarchy.

Architecture, from this perspective, is primarily a technique of mediation between *herrschaftswissen* (governing knowledge) and everyday practices. In the realization of a built form, the meaning and effects of interests change because architecture reacts to them or integrates them, as it responds to the pressures of appropriation and resistance. The practice of participation actively takes up these negotiating functions of architecture.

PARTICIPATION, PACIFICATION AND EMANCIPATION

Participation means taking part in something already given and can refer to any area of social life, work, politics, school, cultural production or consumption. Participation can mean arranging breaks autonomously at your workplace or receiving a share of the company's profits, taking part in a referendum or going to a parents' evening at school, manipulating an interactive artwork, painting your sneakers, or fixing your bicycle or cabinet yourself. You take part in something that would happen anyway, but with your involvement this activity might become simpler, more satisfactory, more beautiful, cheaper, faster, more just, or even more humane and more democratic. Through this involvement, you develop a defined relationship to a specific power constellation, which compels you to participate or at least puts forth the possibility of participation. Such arrangements reveal their borders and structure hierarchies within which you can participate.

But today's more frequent calls for participation can be understood as a new governing principle under which privatization and the call for more self-government legitimize a renunciation of social responsibility. While in the 1960s, demands for democratization and liberalization still had an emancipatory character, at present one wonders to what extent the techniques of mandatory flexibility correspond to neoliberal ideologies or at least to the logic of administration, management and production.

The discursive, process-oriented method of participation is considered economically superior to more hierarchical and static models of organization. Theories of self-organization in biology, mathematics, computer science, economics and urban planning recognized early the benefits of self-correction and the innovative capacities of self-organized contexts. This process of creating less hierarchical social structures through participation is often couched in terms of efficiency and better control. Subjects are addressed as entrepreneurs, and asked as "active citizens" to take their fate into their own hands. The resulting internalization of constraints and self-discipline shifts forms of control, repression, exploitation and exclusion onto the subjects themselves: one voluntarily goes on patrol or tries one's luck as a so-called me-inc. ("Ich-AG")[1].

---

1 Expression for a small one-person-company that was developed recently by the German Federal Government.

Productive debates about participation name these control mechanisms, and suggest social models and cultural techniques for mediating power. Participation, even if it is regarded as pacification or subtle involvement, entails an articulation of political conditions. With luck, this sets something into motion. The resulting conflicts, differences and pluralities refer to an imagined promise of a radical-democratic, equalitarian, pluralistic society and present the possibility of an unfurling of desire.

PARTICIPATION IN SPACE

The relationships of power and space are distinctly drawn. Space is socially produced and is itself the place of production and reproduction of society. Therefore space, especially built space, is both an expression of power and the site of its negotiation. Though space is essentially produced by all who take part in its construction and use, economic, political and social power relations settle there, excluding a multiplicity of possible uses. Housing conditions—whether a single family home with garage, a renovated historic building or former public housing—embody hegemonic concepts of society which reproduce themselves in those spaces as built reality.

Critically understood, participation is an involvement in the social sphere, one that reflects upon society's grounding conditions and possibly steps beyond them. Questions of private property arise, and the economic, political and social criteria governing use. Worker housing projects, Soviet commune houses, people's parks and the public housing complexes of the post-war period were architectural responses to the demands of new social forms. Social movements, as well as the reformers and planners they pressured, called for new spaces. Through such means, resistances were pacified but new social realities were established as well.

These demands for democratization and self-determination, which were common in western democracies toward the end of the 1960s, had a profound effect on debates in architecture. The authority of planners and their role as caring experts stood at the center of the conflict. The architect, legitimized by his profession as a mediator of interests, was confronted with the failure of allegedly objective and technical decision-making procedures, and criticized for his self-understanding as a representative expert in design. The contradictions between authoritarian aesthetics, economic-technical requirements and social reality as well as the perceptions of the users took center stage in his profession. Practical constraints, personal genius and technocratic culture were all called into question. This imposed self-reflection enabled a fundamental questioning of the function of architecture within a social context and fostered new approaches to design.

BUILDING HISTORY

From the beginning, the relationship of western modern architecture and its universalist emancipatory claims to the complexities of lived experience was complicated. There was a fundamental opposition between attempts to best serve the masses through standardization and the cultural promise of the unrestricted development of the individual. This was further aggravated by social, cultural and economic restructuring following the Second World War.

Even the "heroes" of classic modernism had to grapple with the problem of different lived actualities and attempted to find "solutions". The most widespread attempt—the normalization of space and its uses—was only one of many concepts. "Die Wohnung für das Existenzminimum"

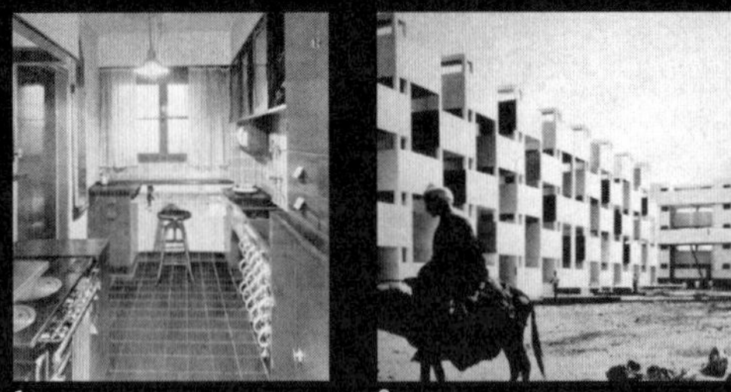

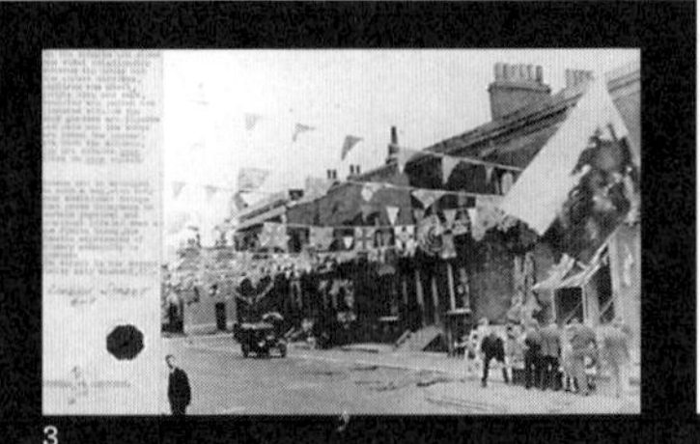

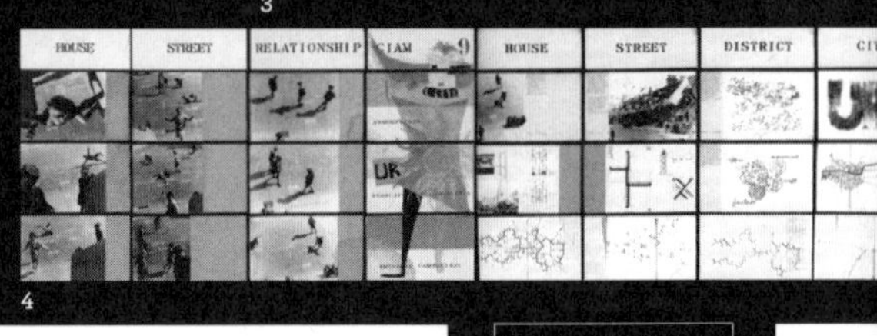

1 Schütte-Lihotzky: *Frankfurter Küche,*
1927–1928

2 Candilis & Woods: *Nid dÁbeille,*
Casablanca 1952–1953

3 & 4 Smithsons:
Urban Re-indentification Grid, 1953

5 Rudowsky: *Architect Without Architects.*
A Short Introduction to Non-Pedigreed
Architecture, New York 1999

(the apartment for subsistence living) and the "Frankfurter Küche"
(Frankfurt kitchen) represent this prevailing tendency, but diminished as it was by construction industry-based functionalism, it was
harshly criticized in the post-war period.

So-called Functionalist Architecture at the beginning of the 20th century had not proclaimed itself an anti-ornamental movement, as was later insinuated by the closely related "International
Style". Their claim was a more general one concerning the effects of built space on human life. In
relation to social realities, there has seldom been a design theory that is more functional than
participative architecture.

From the post-war period until the late 1980s, different concepts of participative architecture
were developed. A widespread approach was marked by a focus on everyday life and regional,
self-organized, or so-called, vernacular architecture. A second line of thought focused on flexibility of the built environment. A third approach emphasized more open, not entirely prearranged spaces. All of these techniques of participation, which were taken up in projects of the
1970s and then further developed, had their beginnings in early modernism. As a rule, these
different strategies for approaching the dynamics and complexities of reality and everyday life
emerged in combination.

EVERYDAY, REGIONAL, SELF-ORGANIZED WITHIN BUILDING HISTORY
Looking back on historic building forms in far-away regions that are credited with fostering a
more direct, original way of living, or even on the daily world of consumption in the west, lends
insight into habitation and its self-organized forms. The vernacular architects assumed that
here—and not in the history of the academic or technical development of architecture—lay the
key to building methods that could embrace the lives and wishes of the people. The evolutionary
character of historical buildings, the self-organized constructions developed in accordance with
circumstances of everyday life, as well as the daily practical testing of building forms were the
foundations of this understanding. They attempted to focus solely on the functions and needs
of daily life, and from this point of view questioned the planner's approach.

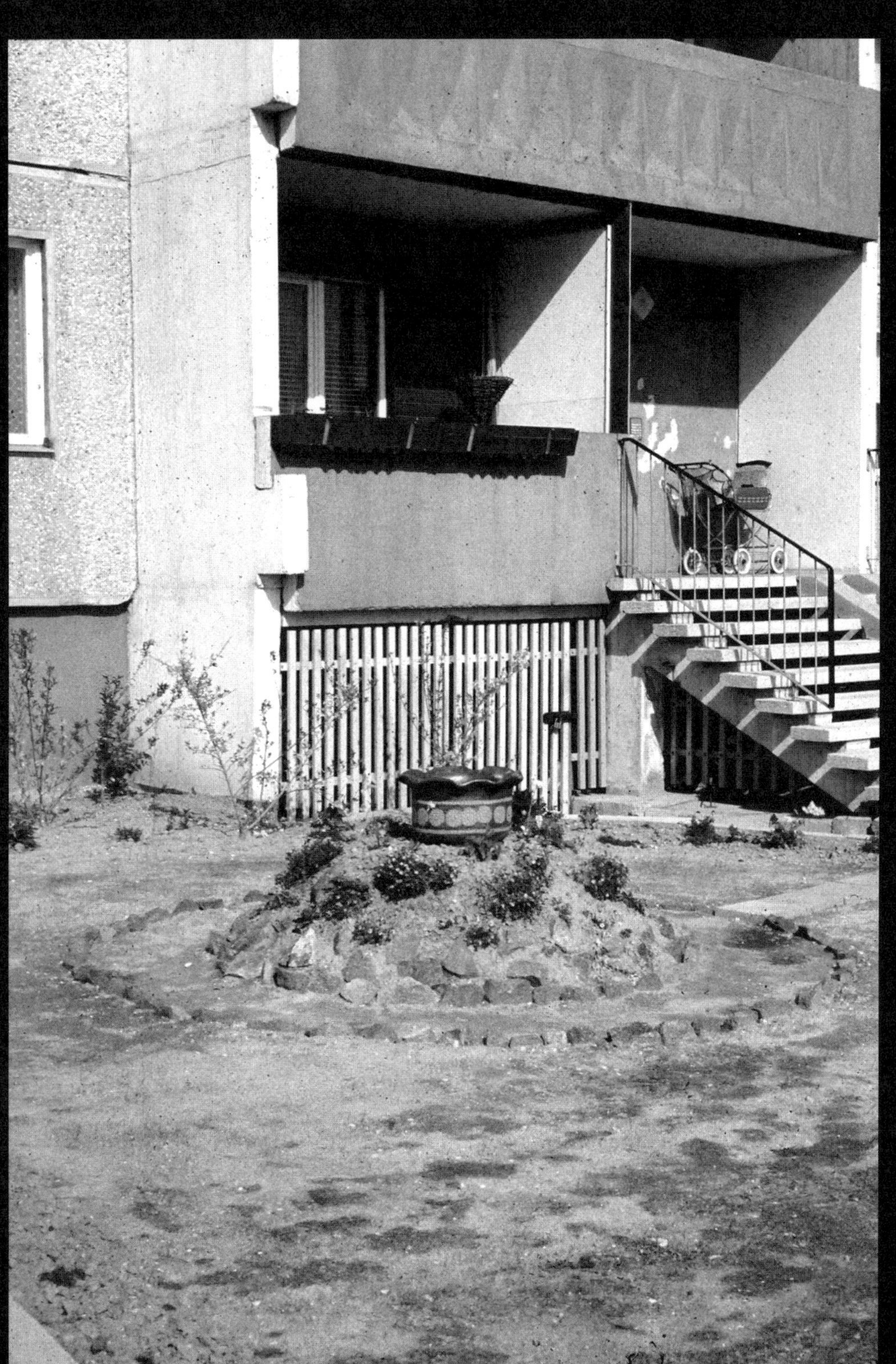

At the CIAM (International Congress of Modern Architecture) congresses after the Second World War, associates of *Team X* turned their attention to the settlements and buildings of the "third world" more intensely than previous tendencies toward exoticism had allowed. Georges Candilis and Shadrach Woods created a block of flats in Morocco in 1953 that transformed regional courtyard types into multiple story residential buildings. At the same time, Alison and Peter Smithson focused on everyday culture in Great Britain. In their "Urban Re-Identification Grid" for CIAM 9 in 1953, they used Nigel Henderson photographs of a street fair and the children's use of the city in the worker and migrant district of Bethnal Green in London to demonstrate their idea of the street as an extension of the home. In the Dutch magazine "Forum," Aldo van Eyck presented examinations of African and South American building forms. A popularization of these perspectives was presented in Paul Rudowsky's exhibition "Architecture Without Architects" at the New York Museum of Modern Art in 1963, which showed photographs of settlements and buildings worldwide that had been built without the help of professional planners.

FLEXIBLE SYSTEMS WITHIN BUILDING HISTORY

The best-known strategy for dealing with the problem of unpredictable use was the creation of flexible space or the mobilization of spatial elements. It was an extension of the functionalist attempt to problem solve; certain parameters of the design became technically flexible and could be integrated according to different spatial demands. The space was still organized by the planners, but the possibility of certain changes was pre-structured.

In 1922 Walter Gropius developed a system of standardized components, which could be combined into a variety of building structures. They could be assembled according to a diversity of living structures and demands. With his "Big Size Construction Kit" ("Baukasten im Großen"), it was possible to produce single-family houses by combining individual components. The "Rietveld Schröder House," built in 1924, accommodated changes in the building's use over the course of day and night, over the course of years and over the course of the inhabitants' lives, as well as the needs of each family member. Gerrit

1 Gropius: *Baukasten in Großen* 1922

2 & 3 Rietveld & Schöder: *Rietveld-Schöder-House*, Utrecht 1924

4 – 6 Le Corbusier: *Unite d' Habitation*, Marseille 1947–1952

7 Archigram: *Plug in City*, 1964

8 – 10 Mies van der Rohe: Apartment House (Werkbund housing estate), Stuttgart 1927

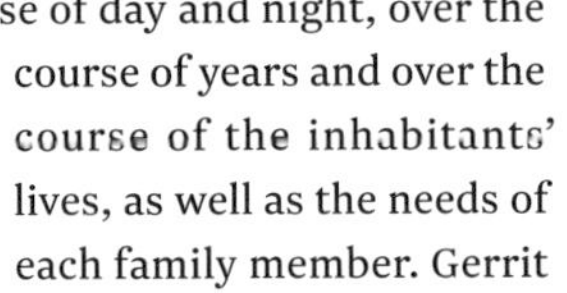

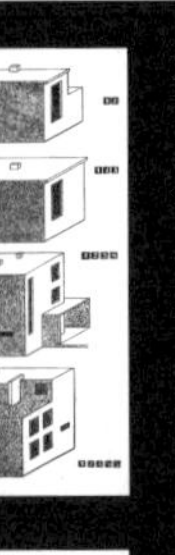

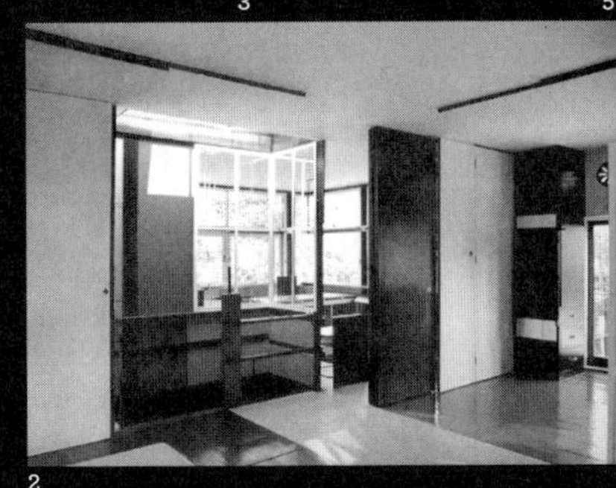

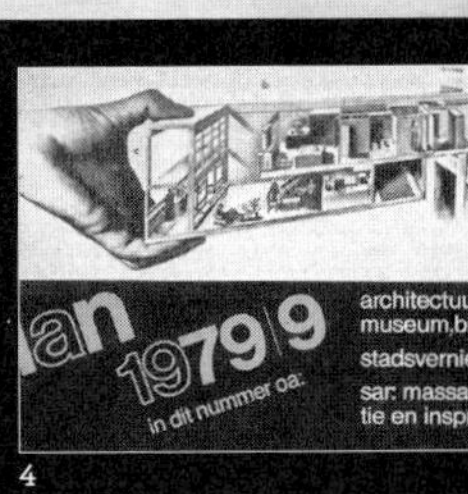

Rietveld installed a system of sliding and foldable walls as well as built-in-furniture, which made the division and combination of different functional spaces possible.

A further iteration was elicited by Martin Wagner's competition "The Growing House," sponsored in 1931 by the Berlin City Council. The study group, which consisted of Gropius, Taut, Scharoun, Häring, Mebes, Mendelssohn, Hilberseimer, Poelzig and Eiermann among others, called for "Bauen auf Stottern" (translated as building on stuttering, which is a German expression that means to pay by installment). They planned a system around an affordable core building that covered basic needs and could be extended and upgraded gradually according to financial opportunities. But it was Le Corbusier who created the most explicit image of flexibility for residential buildings: a photograph showing a hand plugging a living Unit into a lattice structure. This visual interpretation of the "Unite d' Habitation" correlated with the notion of the living cell, the private "Living-Unit" independently integrated into a collective structure. In Japan, beginning in 1958, the architecture group "Metabolists" developed different projects with interchangeable, industrially prefabricated spaces and spatial elements that were finally realized as prototypes for the 1970 EXPO in Osaka. Another prominent example of flexible space was the "Plug in City" by Archigram, in which private spaces were rarely preplanned and were to be created out of prefabricated elements or self-built. In the early 1970s, the German Federal Building Ministry took up the idea and announced the architectural competitions "Flexible Floor plans," "Elementa 72" and "Integra."

OPEN SPACES WITHIN BUILDING HISTORY

Ludwig Mies van der Rohe refrained from drawing floor plans for the apartments in the famous four-story apartment house he built for the Weissenhof Werkbundausstellung 1927 in Stuttgart. The newly developed steel-skeleton construction made it possible to separate the bearing structure from the elements that defined the spaces, so that only staircases, kitchens and bathrooms were fixed. His starting point was the idea that certain areas cannot be planned by experts and that these open spaces could be laid out more effectively and more functionally by the inhabitants or by people working with them. The idea of the open floor plan, which was implemented by Mies and his colleagues otherwise only in luxurious villas, was illustrated diagrammatically by LeCorbusier in his "Plan Obus" in Algiers in 1931. His draft of a strip-city with an integrated highway offered empty levels within which the inhabitants could build their houses according to their own taste. Constant Nieuwenhuys, Yona Friedman

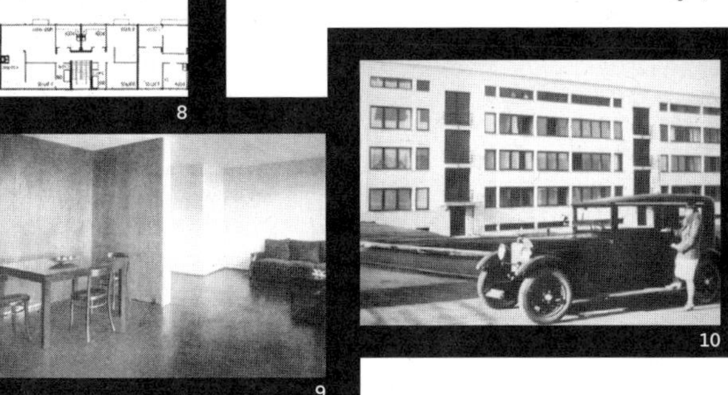

JESKO FEZER AND MATHIAS HEYDEN

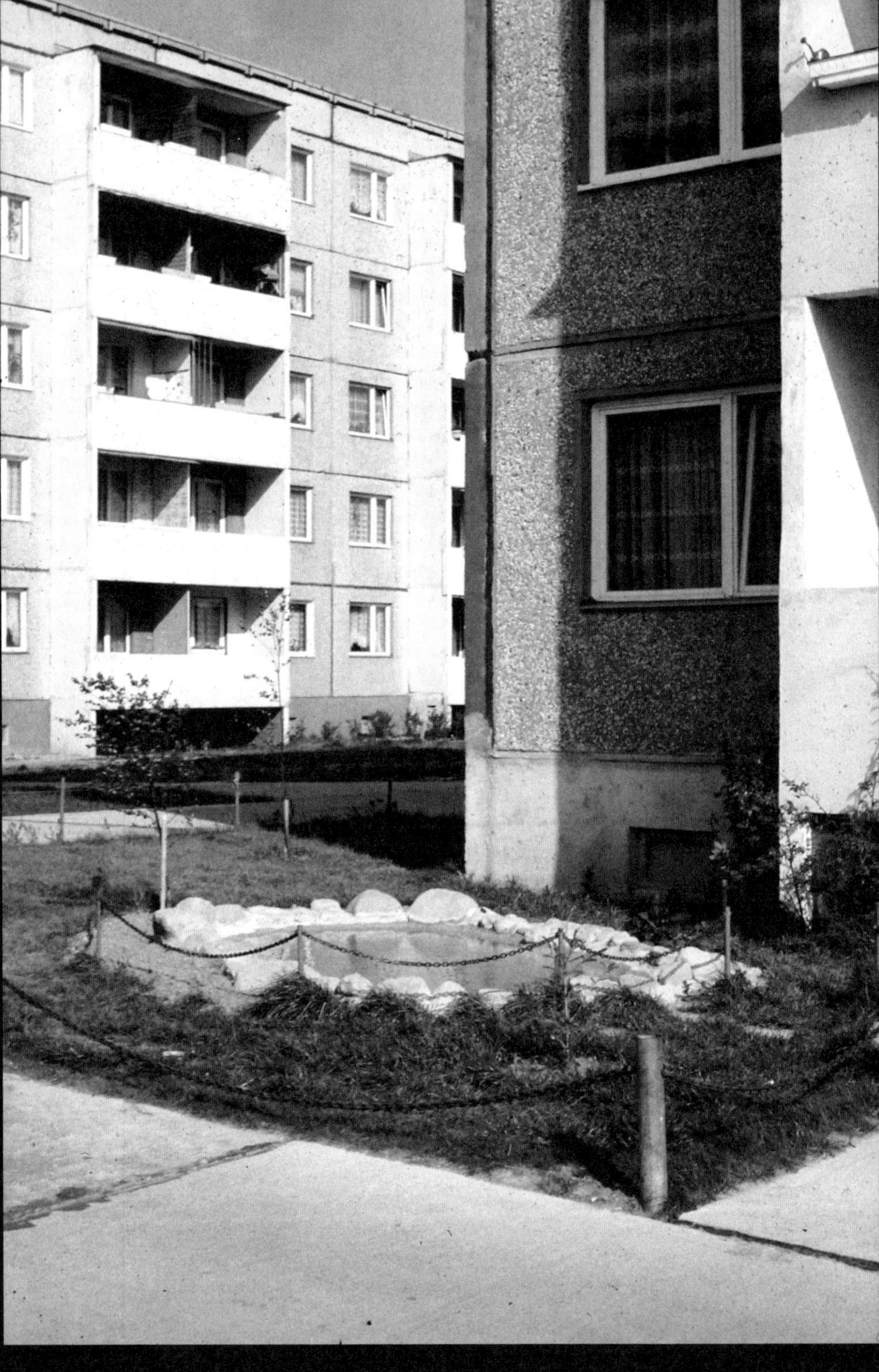

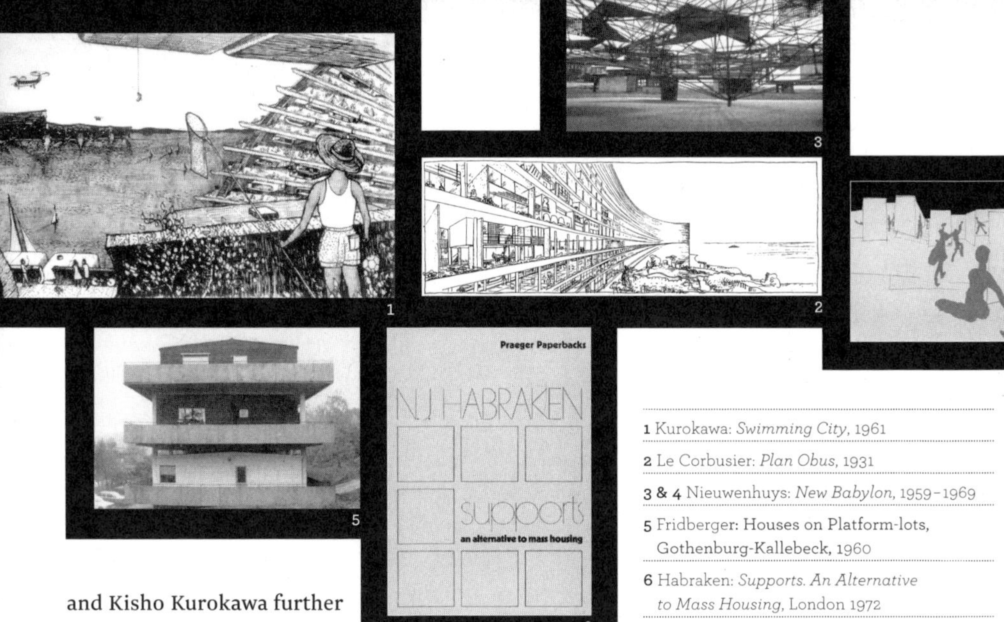

1 Kurokawa: *Swimming City*, 1961

2 Le Corbusier: *Plan Obus*, 1931

3 & 4 Nieuwenhuys: *New Babylon*, 1959–1969

5 Fridberger: Houses on Platform-lots, Gothenburg-Kallebeck, 1960

6 Habraken: *Supports. An Alternative to Mass Housing*, London 1972

and Kisho Kurokawa further elaborate on the theoretical principles of open fields. The first implementation of this idea was a multi-story building in the late fifties by Erik Fridberger in Gothenburg. Nikolaas John Habraken, in his 1961 book, *Supports*, describes in detail the advantages of such a division between a collective support-structure and private development on the resulting floors.

These early attempts to deal with the unpredictability of use in integrating inhabitants continued in the participation projects of the seventies and eighties. In connection with the radical social changes of that time, different architectural concepts were developed: radical-democratic, anarchistic, alternative, techno-utopian, reformist, autonomous, self-help, flexible prefabrication. Given the fundamental planning theory that space and what it contains influence one another, it followed that no space could be built for a presupposed use without those uses changing. Concepts that enable feedback processes, accept failure, privilege the creation of possibilities over finished solutions, work in a process-oriented manner, offer flexible space concepts and finally enable the participation of different people, all refer to these experiences.

GLOBAL NON-PLAN

Under present shifting circumstances, the question again seems pressing: how can we deal with the built environment, given the limits of planning? Procuring land and building one's own four walls are the dominant building practice in large parts of the world, while state planning and regulations often limit themselves to the subsequent formalization of housing projects. Informal markets and other poverty-economies, usually attributed to "third world cities", have existed for some time in prosperous regions of Europe and North America. For example, on the periphery of Rome, more than 800,000 immigrants live in buildings that were illegally established and Athens, despite numerous interventions by government planners, continues to change and expand in an increasingly unregulated and process-oriented manner. Globalized economies and ensuing waves of migration produce metropolises worldwide that are beyond the bounds of planning. The settlement policies of global corporations for their administration facilities and production plants as well as the dynamics of poverty migration largely evade local or national

JESKO FEZER AND MATHIAS HEYDEN

control. These forces of spatial self-organization and the unstable economic and social conditions call the efficiency of conventional administrative instruments radically into question.

The diverse practices of spatial production continuously escape the analytic view. It is increasingly accepted that the conditions under which cities and living spaces evolve and the demands that are made on them are too concealed, complex and mobile for the idea of temporal-spatial controls and fixed space to be maintained. The various strategies of self-organization and models of participative architecture demonstrate ways for dealing with this unpredictability and for accessing different social fields.

WHAT TO DO?
In the context of these altered social conditions, a simple update of western European participation strategies from the 1960s to the 1980s seems unthinkable. The options for dealing with these projects dwindle together with the dissolution of the welfare state to which they were bound. Thus, the end of public housing programs and the breakdown of other forms of state support whose aim was social balance, has weakened the framework for participation projects. Additionally, the forms of participation that were incorporated into building and town-planning legislation have become increasingly ineffectual in the context of privatized urban development. There, participation seems too rigid to be able to effectively carry out alternatives and at the same time too malleable to raise powerful opposition. It is bound to delay procedures by raising objections and is discredited as a way of gaining acceptance from those affected by controversial projects. In the other large arena for historic participation projects—the design of single-family homes in suburban neighborhoods—manufacturers of prefabricated houses have taken over the professional production of wishes and fulfilment of desires. In this context, the role of the planner, as it was developed in the context of industrial capitalism, is greatly reduced.

These changes take place in the context of overlapping nation-states and global economic alliances, as well as in relation to new social alignments and their intensification and acceleration of exchange. These multi-layered networks and the self-determination that is demanded from and gained through them, pose new challenges for spatial design and practice. It establishes local spaces for different groups in which community, culture, alternative economies and life-styles are lived out and tested. Spatial production grounded in involvement enables people to partake and to negotiate with others and thus makes social plurality productive. Participative architecture questions the organization and use of space and exposes existing social relations within an ongoing negotiation process.

In this context, the architect who is integrated into very diverse communication hierarchies is well suited to take up new tasks. Experienced in multidisciplinary discussions and negotiations with owners, users, construction companies, contractors, craftsmen, engineers, investors or administrative authorities, architects seem best qualified to take up the varied challenges and forms of participation. Participative architecture searches for new alliances, working methods and methods for spatial organization, for approaches that accommodate multifaceted and changing life-styles/ make living in the plural possible / take up different ideas of the use of space // reveal economic conditions for self-determination / call into question real estate property / put forward a collective understanding of ownership // accept the dynamics of self-organization / use the potential of self-building // design in an interdisciplinary way / develop spaces that are not fixed to a final state / react to future demands / contemplate parametric architectures / program changeable rule sets / structure diversity with variants // demand social interaction / open structures of decision-making / extend the temporal and personal spaces of negotiation / assume an unlimited multiplicity of participants / assemble communication techniques for non-professionals / make relevant information available / initiate public debates on the built environment // understand planning as making things possible / offer technical support / initiate negotiation processes through provocation / question adjustments and competence relations / lend inspiration for possible developments by outlining scenarios / evaluate possible consequences / offer use-neutral spaces / take into account possible misuses and temporary appropriations // look for tasks that users offer / think of a building on demand / offer planning competence to economically excluded people / look for contact with the culturally excluded / stimulate the spatial production of social structures / invent collective planning tools / accept failings and compromises / promote self-generated aesthetics / overcome fixation on images / use the communication tools of information technology / apply flexible constructions / test production methods of mass-customization / integrate flexible parameters in built structures.

INTERVIEWS

The interviews and lexicon included in this project and summarized here aim to critically examine the history and theory of participative architecture as well as recent applications, so that fresh perspectives on participation and self-organization might uncover new possibilities for collective appropriations of space.

One possible starting point is an analysis of *spatial appropriations* that take root outside official planning processes and the ways in which they relate to a given place. The architect, *Wolfgang Kil* retrospectively analyzes unplanned, but tolerated appropriations in GDR social housing complexes, using the example of re-designed gardens and balconies. The structural rigidity of the buildings served as a framework for appropriation through informal renovations and decorations: The collectively owned dwelling was something to be taken into possession, rather than a commodity with obligations to preserve its value. On the city level, architect *Yvonne P. Doderer* refers to realms of emancipation and socio-political resistance within the disciplining logic of urban planning. Using the appropriation of an inner-city area by womens' project groups as an example, she develops a feminist perspective on the potential of the urban. Against this backdrop, she calls for the tolerance and promotion of independent initiatives against procedures of control and the moderation of participation. The architect *Eckhardt Ribbeck* departs from the European framework by taking up the historical observations of unplanned appropriations of space by John F. Turner and Colin Ward in the 1970s. Taking Mexico City as an example, he describes illegal self-building as an effective production of space under conditions of extreme poverty and the absence of national regulations. Conditioned on an extremely capitalistic property market and effectively simple construction methods, a fixed structure develops that is open to spatial variety and constant alterations. This form of involuntary self-activation under

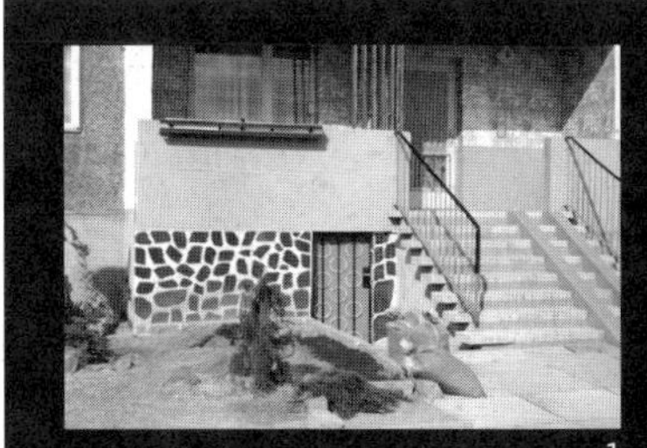

1 Kil: Wolfen-Nord mid-1980s
(also see images on pages 10, 15, 18 & 22)

2 Doderer: *Urbane Praktiken* (Urban Practices), Münster 2002

21

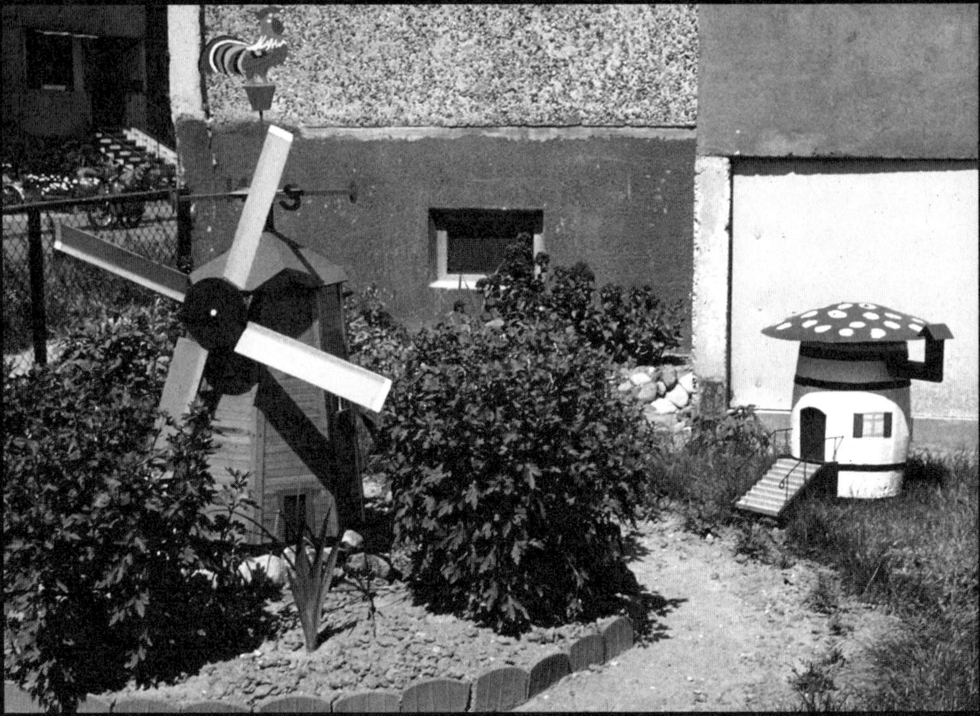

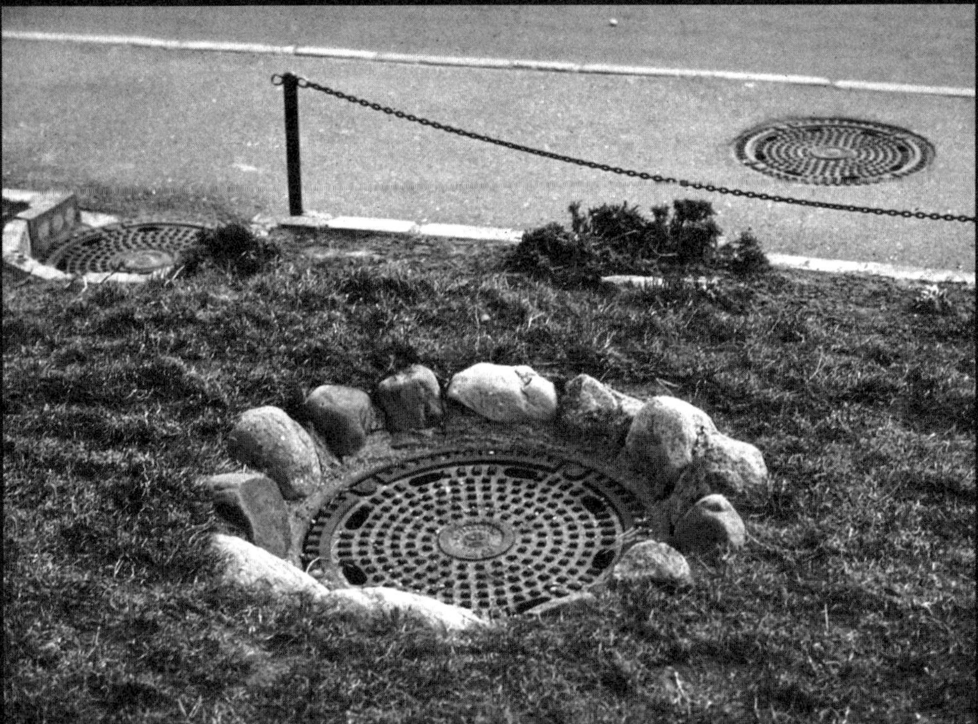

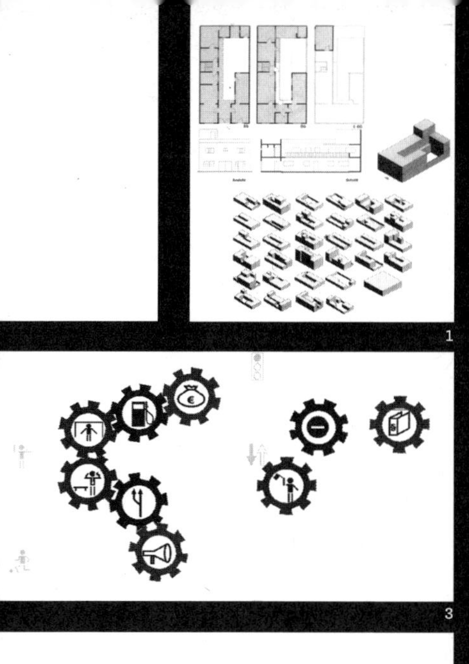

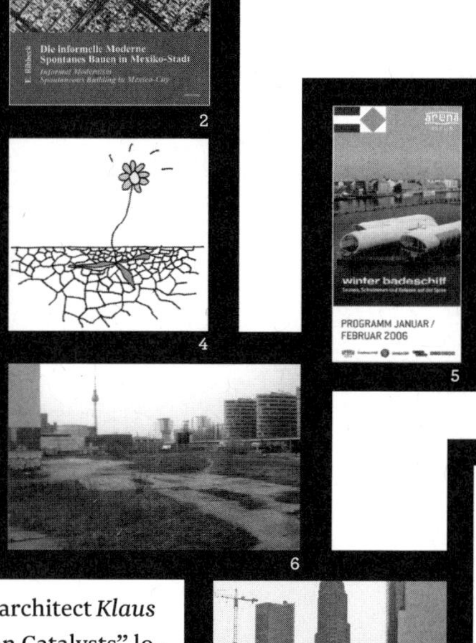

economic pressure is the only possible means of attaining space necessary for survival in such regions. Within the phenomena of intermediate uses, the landscape architect *Klaus Overmeyer* from the research project "Urban Catalysts" locates the spontaneous and unplanned in temporary spaces. These spaces lie outside the attention of planners and the pressures of use resulting in different processes of appropriation. In dealing with these new requirements, planners position themselves as potential "enablers". New tools for process-oriented planning must be developed and new participants included. This process of formalizing intermediate uses must also always battle with more controlled claims on space and with the anticipated profits of real estate.

Experience of self-organized productions of space generate different *criticisms of planning*. Using examples of self-organized structures in Europe, the architect *John Palmesino*, part of the spatial research project "Multiplicity", criticizes the analytical planning perspectives on territories and subjects which assume that the acquisition of rational knowledge is a basis for design. He challenges the planner's objectivistic perspective with specific local structures being developed without plan by a multiplicity of participants. The project "Divercity", of the Berlin-based architecture office *ifau*, argues against the simulation of variety in urban planning and the immobility of fully regulated planning tools, in favor of a complete withdrawal of rules in order to generate activity and encourage co-operation. With open source programming as a model of collectively structuring an open design process, ifau describes how one can use access to information, provision of communication tools, and flexible planning instruments to spur participation of different participants and start an open-ended process of town planning and

1 & 2 Ribbeck: *Informal Modernism. Spontaneous Building in Mexico-City,* Heidelberg 2002

3 – 7 Images promoting vacant sites and strategies for "Urban Pioneering"

8 Palmesino (Multiplicity): Multiplicity, *USE—Uncertain States of Europe,* Milano 2003

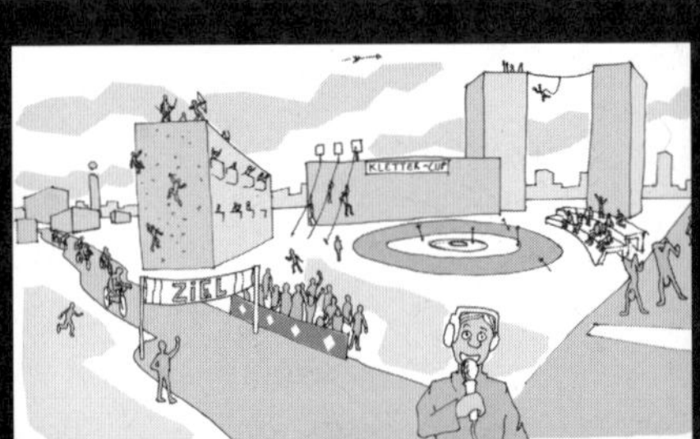

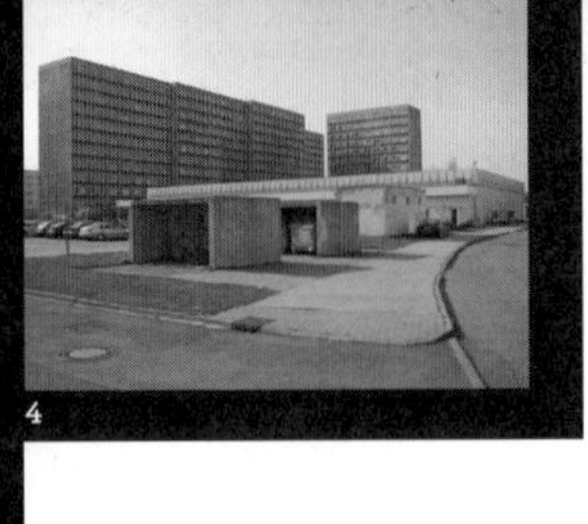

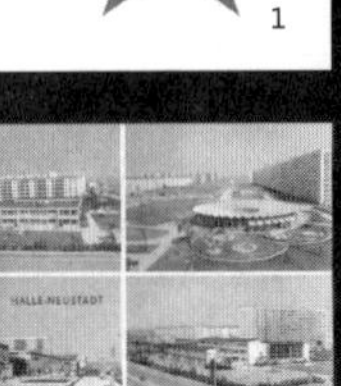

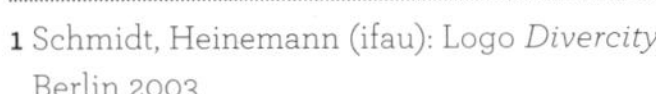

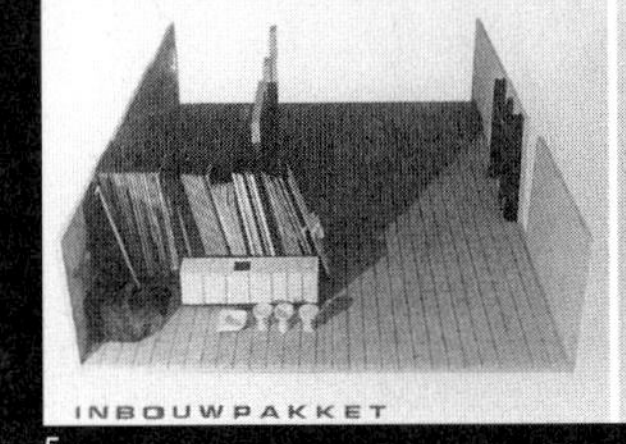

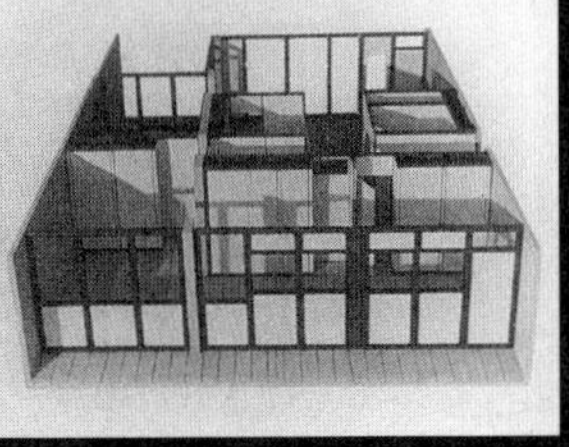

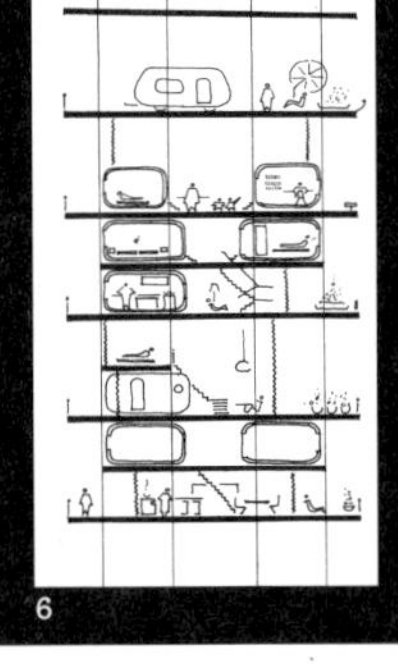

1 Schmidt, Heinemann (ifau): Logo *Divercity*, Berlin 2003

2 – 4 Bader, Forster-Baldenius (raumlabor-berlin): Halle-Neustadt in the 1960s and 2003, and illustration of *Kolorado. Perspectives for Halle-Neustadt*, 2004

5 & 6 Habraken: Section-scheme and models according to the concept of support and infill, Netherlands 1960s

7 & 8 Loeper: Apartment building (support) and open-floor-plan-flat (infill), Berlin 1970s

9 & 10 Friedman: *Ville Spatial* 1964, and Sabine Lebesque, Helene Fentener van Vlissingen, *Structures Serving the Unpredictable*, Rotterdam 1999

11 Fritz: Scheme on top-down and bottom-up planning-techniques, Kaiserlautern/ Rotterdam/Zürich 2002

12 & 13 Kraft: Model of a growing house, Berlin 2003

architecture. Another strategy for the diversification and integration of diverse players is being developed at present by the architectural group *Raumlabor_berlin* for Halle Neustadt. Diversified participation strategies should impact the development of neighborhood identities within the city. *Raumlabor_berlin*, recognizing the limitations of a master plan, offers a spatially partitioned, process-based planning model, which develops communication structures, scenarios and tools.

Another strategy to deal with the limitations of planning is to activate the potential of *industrial building*. In the Netherlands, *Nicolaas John Habraken* dealt early on with the possibilities of individual user participation in the industrial construction of houses. The Dutch research project "S.A.R.," founded by Habraken in 1964, located the problem of massive housing projects of the post-war period not in industrialization and economics, but in the enforced elimination of the inhabitant from the building and planning process. Arguing against a paternalistic conception of architecture, he calls for a redefinition of the role of the planner in favour of real industrial manufacturing and user participation. He suggests a principle of open planning based on spatial considerations that are not fixed on final forms and makes those accessible and communicable. There is also the model project "Variables Wohnen" (variable dwelling) of the Bauakademie in Berlin and Rostock, one of the few building experiments in the GDR, which was based on active user participation and concerned itself with the potential of open-ended industrial construction methods. *Herwig Loeper*, architect, was involved in the planning and realization of this structurally and conceptually significant project. He reports how in the early 70's in the manufacturing of apartment buildings in the GDR, the parameters of standardization were tested

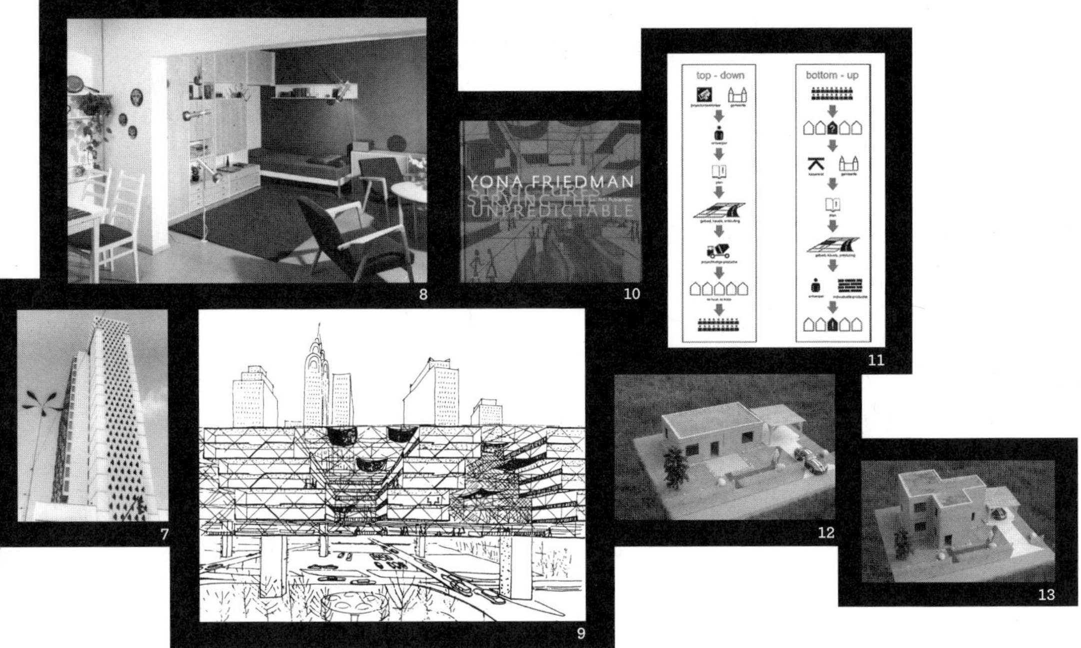

experimentally. Parallel to Western European experiments, the introduction of flexibility and adaptability of apartments problematized concepts of regularization and self-determination, and thereby set forth possibilities for individualization within the conditions of state socialism. The architect *Oliver Fritz* describes the principle of mass customization and recent approaches to component configuration through variable parameters instead of fixed end forms. With plans as modifiable data files and individualized production, urban planning and architecture according to the differentiated desires of users appears to be technologically possible. The engineer *Udo Kraft* is presently developing in his project, "the growing house" ("Das mitwachsende Haus"), an economic and pragmatic construction model for the cultivation of affordable and modifiable living in the private home. Kraft demonstrates how cost-efficient building must take place gradually in order to remain flexible and therefore sustainable in the future. These structurally realistic projects still stand in opposition to the German "my own home"-mentality, the architect's self-understanding and the economic logic of the building industry.

An additional model of handling the limitations of planning is *self-building* by the users. In the late 1950s the Hungarian-French architect *Yona Friedman* developed an urban-design principle of open primary structures for free and self-determined use. He promoted development of flexible spaces and the potential for do-it-yourself building methods. To empower the inhabitants, he used simple sketches as communication tools and tested his theories with realizations like his self-building projects in India. The self-determined development of space is described by the energy technician, *Martin Stengel* in his account of "Ökodorf Sieben Linden," a contemporary, live -work collective. This project for an alternative model of society, which understands itself to be a laboratory situation outside—urban everyday life and its obligations, also requires architectural-spatial self-organization. Thus, the communication of direct democracy and social self-authorization are reflected in the do-it-yourself-constructions of the settlement.

These different forms of engagement with users always refer to a *collective project* of self-determination and self-organization. A mostly ignored instrument for enabling other building and living forms lies, for the economist *George Knacke,* in the principle of the co-operative. The

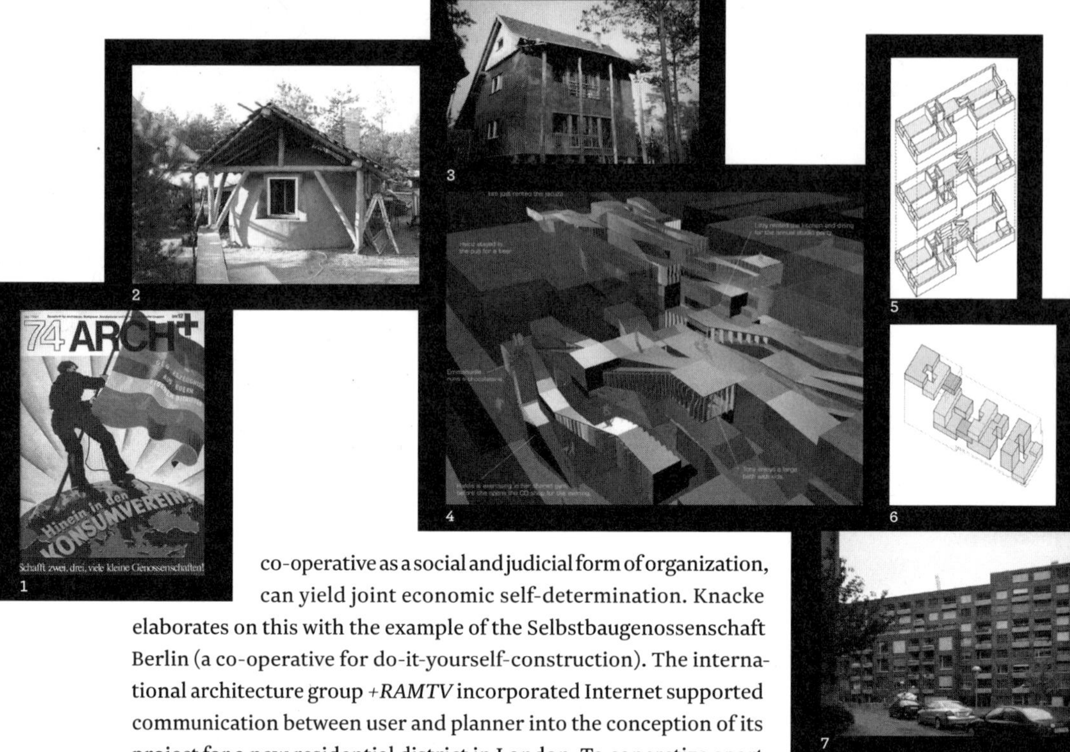

co-operative as a social and judicial form of organization, can yield joint economic self-determination. Knacke elaborates on this with the example of the Selbstbaugenossenschaft Berlin (a co-operative for do-it-yourself-construction). The international architecture group *+RAMTV* incorporated Internet supported communication between user and planner into the conception of its project for a new residential district in London. To concretize apartment prototypes, the wishes of potential users were collected and organized in relation to one other. This tool for overseeing participation is also seen as a way to strengthen interaction amongst the inhabitants prior to the building phase; in order, for example, to structure the relationships within private areas so that overlapping spatial programs and common use of private spaces were made possible. Beyond the possibilities of more effective communication and computer-assisted variable planning, +RAMTV emphasizes new production methods for the individualization of space. *Andreas Hofer* of the co-operative project "Kraftwerk1," describes the dynamics of the organization and communication structure of self-initiated group living projects in the urban context of Zurich. On professional and diverse volunteer levels, proposals for the building and its use were compiled. At Kraftwerk1, architecture was developed relatively independent of social organization. Spatial co-determination was made possible more through the planning of various sized rooms and configurations than offers to participate.

How *different the role of the architect appears* in connection with self-determined planning and construction processes, was already apparent in the early debates about the possibilities of opening planning concepts and the architectural theory of the pioneers of a participative architecture. The Belgian architect *Lucien Kroll* juxtaposes the still existing Fordist logic of separated and rationalized spaces with implicit user participation. Kroll offers no planning theory and no social techniques for persuasion in his projects, instead he has practiced since the 60s an anarchist-subjective approach. His communicative praxis sets the complexity of the user-subjects against the reductive limitations of industrial building methods. He thereby extends the demand of the users to be part of the building phase across the entire planning and building process. He practices this in the office and on the construction site. In Austria, *Eilfried*

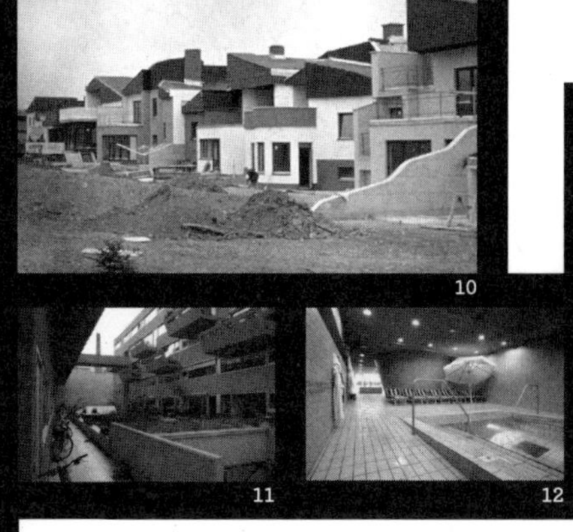

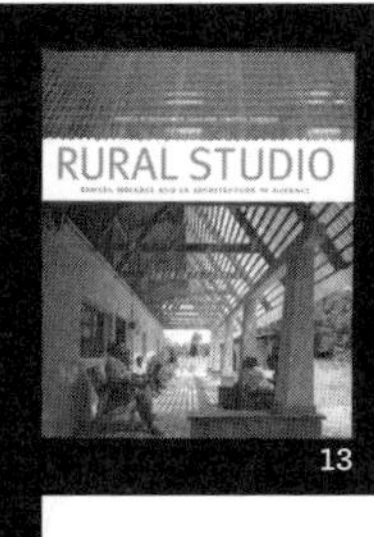

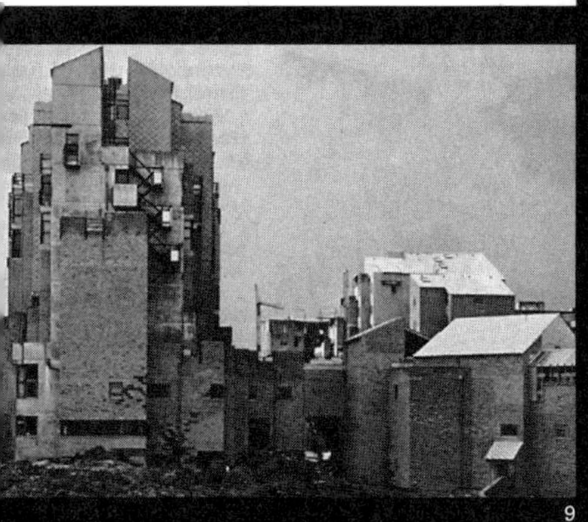

Huth realized radical-participative building very early in the context of local housing support. Extending the role of the architect, he implemented a different kind of architecture on a local-political and building-law level and thus opened possibilities for self-construction, self-planning and self-determined spatial design. *Johnny Winter* of the Viennese office *BKK-3* describes the living projects of "Sargfabrik" (coffin factory) and "Miss Sargfabrik," which he initiated and planned in collaboration with others. Together collective expectations and proposals were developed and expressed in innovative collective rooms, which among other things were also available to the inhabitants of the local neighborhood. Winter stresses that particularly through extensive participation, an ambitious architecture with experimental spatial forms was made possible. *Andrew Freear*, co-director of the "Rural Studio," practices in his architecture training an intensive confrontation of architects with social reality. In a 1:1 building praxis to support underprivileged and destitute persons in construction projects that were not controlled by local building offices, Rural Studio found and donated materials according to the demands and needs of the users. They continue to practice socio-political pragmatism with their building experiments in a clear, demonstrative way, reinventing the role of the architect as a problem solver.

1 Knacke: *Arch+ 74*, Aachen 1974

2 & 3 Stengel (Ökodorf Sieben Linden): Strawbale-buildings of a particular group, Poppau 2002–2004

4 +RAMTV: Illustration of interactive and collective dwelling, London 2002

5 – 7 Hofer (Kraftwerk1): 9-bedroom-apartment, main building and volume-scheme of apartments within, Zürich 1995–2001

8 & 9 Kroll: Student housing, Brussels-Louvain 1968–1979

10 Huth: *Eschensiedlung*, Deutschlandsberg 1971–1992

11 & 12 Winter (BKK-3): A common yard, the common Spa, Wein 1994–1996

13 Freear (Rural Studio): Andrea Oppenheimer Dean and Timothy Hursley, *Rural Studio. Samuel Mockbee and an architecture of decency*, New York 2002

The preceeding text by Jesko Fezer and Mathias Heyden is the introduction to the publication *Hier entsteht. Strategien partizipativer Architektur und räumlicher Aneignung* (bbooks, Berlin, 2004 & 2007), which evolved from an event of the same name held in Berlin in 2003. The event included a scaffolding structure built around and above a 1960s pavilion (adjacent to the Volksbühne theater on Rosa-Luxembourg-Platz), which operated as a space for exhibitions, daily lectures, and screenings as well as serving as an open space for spontaneous settlements and unpredictable activities. Following the event's theme, the publication is a document on strategies of participative architecture and spatial appropriation with a focus on Western Europe. Beginning with projects and discourses from the 1960s, the publication follows this history into present-day concepts and experiments through extensive interviews and related materials. This introduction provides an overview of the themes and theoretical concerns that run throughout the interviews.

JESKO FEZER AND MATHIAS HEYDEN

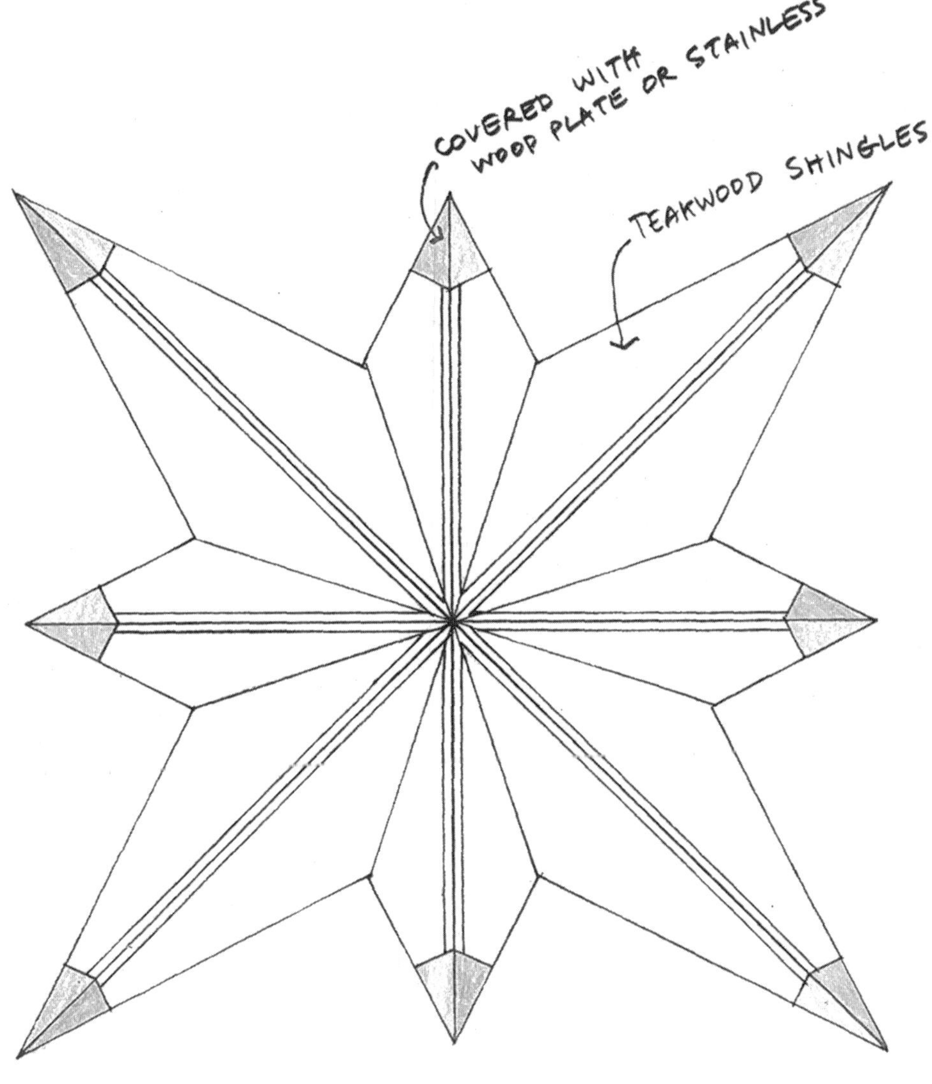

ROOF PLAN 1:50

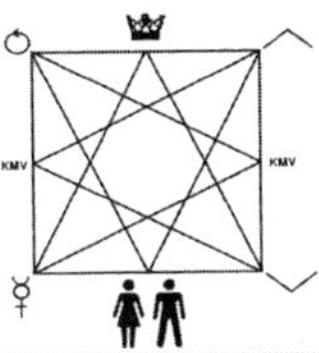

Carl Michael von Hausswolff

KREV
Jürgenson Starhouse, The Land, Chiangmai, Thailand

In summer 2001 I went with my daughters, Anna and Maria, on a holiday trip to Thailand. Apart from spending some time in Bangkok and Koh Chang, we also flew up to Chiangmai together with curator Thomas Nordanstad, clothing designer Filippa Knutsson and their baby son Dylan. As we were making plans for this journey we also talked about visiting The Land, a small piece of land outside Chiangmai where artists Kamin Lertchaiprasert and Rirkrit Tiravanija had created some kind of art community project. Both Thomas and I had known Rirkrit and his work for some years but the actual Land project was something of a legend for us and also…none of us knew Kamin well. The only thing we were aware of was that some other artists had built houses on this Land and some were planning to build more.

As summer in Thailand is the rainy season, the weather on the day we visited The Land was pretty dull. Not cold and grayish dull in a Northern European sense, but warm, wet and cloudy. The heavy rains never really managed to dry up before the next shower appeared. It was a perfect day for a visit. We would be able to see the place without having a wonderful sunny sky misleading us. Leaving our base, a fancy mountain resort, we met up with Kamin Lertchaiprasert at his home and after driving some 20km we were standing in the middle of a rice field surrounded by four or five pavilions, a larger centrally situated building and a few buffalo. It felt like a secret place, a spot chosen for its off-main road qualities, yes, a place for outsiders and mavericks. Listening to Kamin's ideas about the projects—ideas of a non-art site where people could stay and live from what the area produces—I realized that it resembled earlier experiments from the late 1960s and early 1970s. It was similar to semi-autonomous collectives and alternative living situations established in the US and Europe and somehow related to kibbutzes in Israel and kolkhozes in the Soviet Union but was also quite different. First, it had the feel of ambivalence, a vision that included success and failure, which meant that, in practice, this did not have to work. It might as well just rot away, and I realized that this was more of a statement than a refuge from capitalism, drug-abuse and loneliness. It was a couple of artists' physical expression of the idea: This Is Not The World We Want To Live In—We Would Like To Have Something Else! Secondly, and this struck me the most at first, was that it had, from an ideological point of view, a Thai-Buddhist platform where generosity and respect towards life in general rules. I consider Thai people to be one of the most friendly and generous peoples in the world because

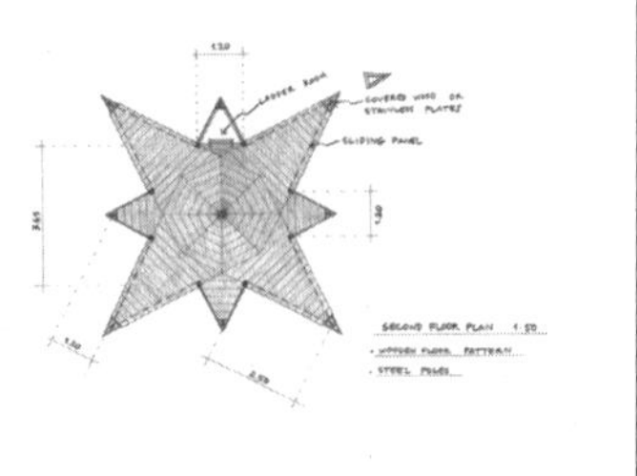

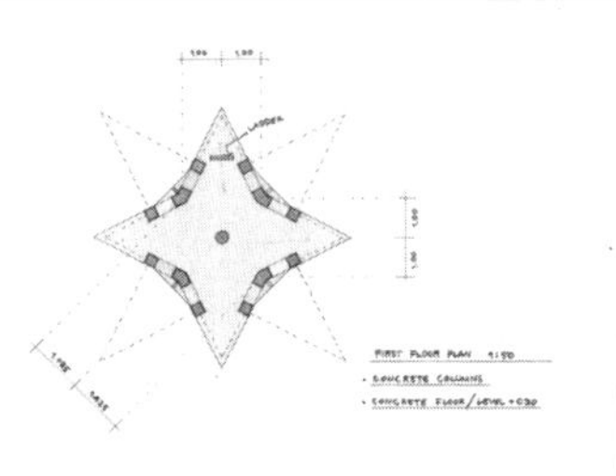

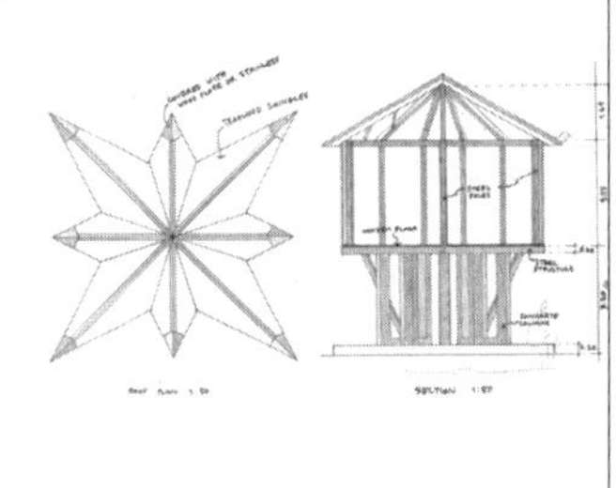

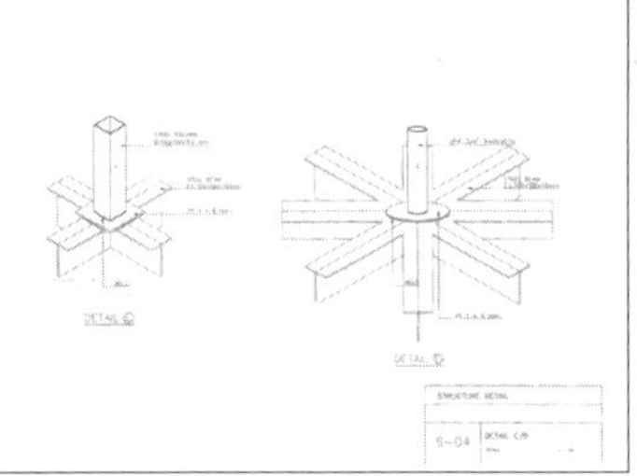

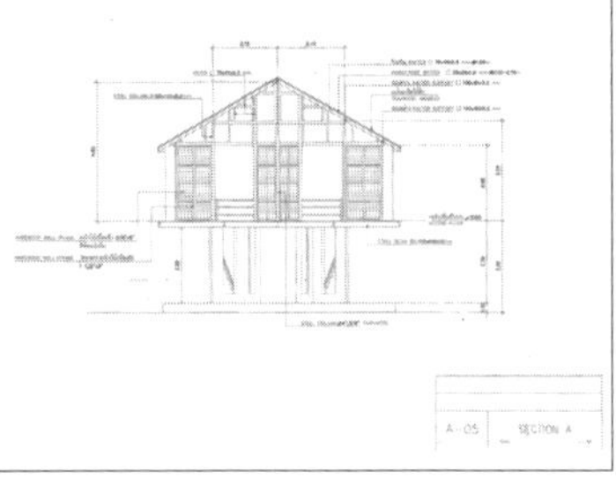

they have been brought up from childhood within this framework of Thai-Buddhism.

After leaving the area we asked Kamin about how they selected and invited artists to participate in the project and learned that it was more of a natural pace factor that made the selection—if one happened to cross paths it was natural for an artist to choose to participate or not. We said to Kamin that it would be interesting to do something like a house. He smiled and said: It's up to you!

Leaving the flatlands for the resort pool we talked about this in the car, excited by the idea of building a house at The Land. One way or another we wanted this to happen.

Back in Stockholm we discussed the subject a few times but we never really got ourselves hanging over a drawing desk and we never even talked about shape, function or size—the collaboration stalled. In 2002 Thomas and I developed the thirty-minute video piece "Hashima, Japan, 2002," a work that felt excellent in its collaborative dimension, but the house project was hardly mentioned. It was not the right time for it— and there was no budget. Personally, though, I could not let go of the idea. Back in Bangkok later in 2001 I got hold of Sumet Jumsai's book "Naga-Cultural Origins in Siam and the West Pacific" (with contributions by R. Buckminster Fuller) partly dealing with traditional Thai architecture, and this reading caused a second rush of inspiration. The use of stilts and the conscious approach to natural and spiritual conditions in both material and form became obvious to me. In October 2003 I met the Thai artist Navin Rawanchaikul on a scouting trip to develop a piece for the 3rd Liverpool biennial. He invited me, within the framework of a larger project called "Fly Me To Another World," to hold a workshop for media art students at the University in Chiangmai in spring 2004, and this was the starting point for me to actually construct the initial drawing plans for the house. Without consulting Thomas, I started to figure out the concept and the form for the building.

My first problem was purely personal. Did I really want to design a house? Why would I want to enter into a project including architecture when an architect who married my mother after she divorced my father wrecked my childhood? Not that I had had any objections to his profession, the problem had been that he was more or less a psychopath, abusing me, my sister and my mother between 1966 and 1976 in a mental and physical way. There had been no lack of wealth but a lack of love. Realizing now, in 2004, that his architecture had been pretty boring and not particularly successful, I decided that he wasn't a threat anymore. He was just a sad memory from the past, living his last days in a remote town somewhere in Denmark. It actually made me happy to be able to think about this person again and leave him there on a dusty shelf in the archive cellar. I also realized that he had given me something that could aid me in this project. His architectural plans and models had been floating around where we lived and when I was thirteen or fourteen he even managed to convince me to help him out a couple of times in his office copying drawings and plans, so I had a feel for this already. Having gotten rid of the darker sides of the past there were new situations coming up: Form!

I knew already that I wanted the building on stilts. All the other houses at The Land were on stilts so this wasn't odd at all. I also knew that I wanted to use material from that part of the world, especially the wood. Very fast I also knew the form of the ground and first floor. It had to be the form of the eight-pointed star that takes shape when the hierarchical diagram of the conceptual state of Elgaland-Vargaland is drawn.

This diagram consists of four triangles put together forming a quadratic shape. This hierarchy was constructed as part of the above mentioned state's constitution presenting another formula for handling the vast problems that occur when a person described as "someone that is a part of the people" meets a person that is described as "The King." This new system equalizes all and everything by rotating and connecting not only humans but also other types of life forms such as land- and air-based animals, valleys and mountains. This work, this state, was created by Leif Elggren and myself in 1991-92 and has since developed into an organism in itself, criticizing the basic fundaments of the national-political world. Its latest embassy was inaugurated in the fall 2006 at the Haus der Kulturen der Welt in Berlin and its latest annexed territory is Isola di San Michele, the island of the dead, outside Venice, Italy (www.elgaland-vargaland.org/).

In May 2004 I had my initial sketches of the house ready and I left Stockholm for Chiangmai. This time I stayed for ten days and my residence was the Land Foundation guestroom on the first floor above the office in a place very near Wat Umong, one of the most remarkable Buddhist temple grounds I've been to. I had three missions there. First, I was going to conduct a workshop based on sound with the students at the university; secondly, I wanted to photograph a new series of "red" pictures based on my interest in abandoned structures and locations; and thirdly, I wanted to know how I could develop the Starhouse at The Land. I met Kamin Lertchaiprasert on my arrival to the office and immediately we made arrangements to meet an architect he had worked with before. The next day I met architect Sawapat "Pat" Chaiyarerk and she turned out to be a superb partner for me. We talked about the structure and I showed her my drawings. We agreed to meet in seven days. She would then have a smaller model built and an idea of how much the whole thing would cost. The same day we also went out with Kamin to visit The Land again to mark out the place where the house should be situated and I also told Pat and Kamin that I wanted a Feng Shui expert to locate the building site more precisely and rotate the structure in order to harmonize it with nature and its magnetic fields. Apart from the structures I had already seen a few years earlier created by Rirkrit, Kamin, Tobias Rehberger, Mit Jai and Superflex, there were also new projects at The Land by Philippe Parreno/François Roche, Angkrit Ajchariyasophon, Thaivijit Puangkasemsomboon & Somyot Hananuntasuk and Kaew. It was all going forward! The second meeting with Sawapat was as smoothly as the first. She showed me the model of the house and for the first time I realized what it would look like. Amusing indeed! She made a remark on the roof and stated that she didn't think that it was a good idea to have that shaped as an eight-pointed star but rather as a normal square shaped roof. This was not from an aesthetic point of view but from a practical one: when the rain season would come the floors would be flooded by water if we used my form. I said I would think about it and notify her via email later on. She also suggested that we should build the stilts out of concrete instead of the large wooden legs that I wanted first, and she also wanted to use steel as a main material for holding up the ceiling- and the floor- structures and to use new Thai wood for walls and floor and old recycled wood for the roof. I said, "Fine, whatever you say!" She also presented a neat figure of what the price tag would be. Leaving Chiangmai after these ten days I realized that I was confronting a new situation: The Financial!

I thought that financing this project would be a piece of cake but it turned in another direction. I was naïve enough to believe that my latest solo exhibition would fill my treasure chest but that show was as disastrous as the earlier ones. Luckily I had stumbled into Peter Weibel, the director at ZKM in Karlsruhe, earlier that spring when I was engaged by the Finnish electronic music duo Pan Sonic to play concerts with them and the Finnish artist and instrument technician/constructor Erkki Kurreniemi. One of the concerts was at ZKM and Weibel came around asking how much these vintage instruments we were playing cost. Realizing that he was some sort of a tech-fetishist fan I drew his attention to Frankfurt am Main where I had just installed the entire sound tape archive recorded by Friedrich Jürgenson. Jürgenson's major achievement in his life was that he was the first person able to record voices "from the other side"—dead people (www.fargfabriken.se/fjf/life.html)! Weibel went to Frankfurt, saw the show and immediately wanted to incorporate the one thousand 1/4 inch tapes into the ZKM collection. I told him that I didn't want to treat Jürgenson's work as a commercial object and suggested that if ZKM could pay the rent for my apartment for one year and the construction of the house at The Land the tape archive would be theirs. Weibel said OK! The money was transferred to Chiangmai in December 2004 and the actual building of the house started. It was set to take around two months if every-thing went as planned. Meanwhile, pictures from the Land were mailed to me showing the Feng Shui expert P Su carefully working on the site.

In February 2005 I was invited by Navin Rawanchaikul again, to participate in the second part of his "Fly Me To Another World" project: this time it was more of a seminar type of program with many other people from the international art world involved. The project circulated around Asian culture versus western, and the local point was the senior Thai artist Inson Wongsam, a marvelous person with a fantastic story to tell. The time was also perfect for me since the house would more or less be standing there. Now it was logically named The KREV Jürgenson Starhouse and I was foaming like a rabid dog to get myself to the Land. It turned out that the entire structure of the house including the concrete fundament, the concrete stilts, the metal frame and the wooden roof was there but not the floor. Getting to The Land together with all the other participants in Navin's seminar was a thrill and I was especially happy that Navin was there with Rirkrit and Kamin and a fourth Thai artist I had known for a few years: Surasi Kusolwong. Seeing the house I could hardly believe that it was actually there. The transformation from the wish to do something to the actual house standing there was breathtaking. It was larger then I anticipated. It was sturdy and strong and at the same time there was a light weightlessness to it. I liked it...very much. Going out there the next day with Sawapat and Kamin to meet the

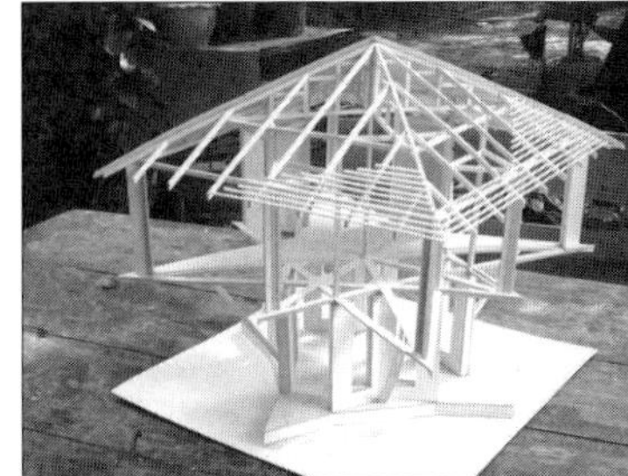

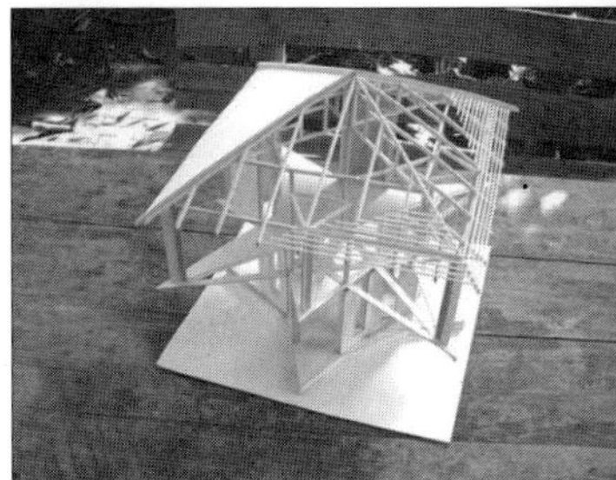

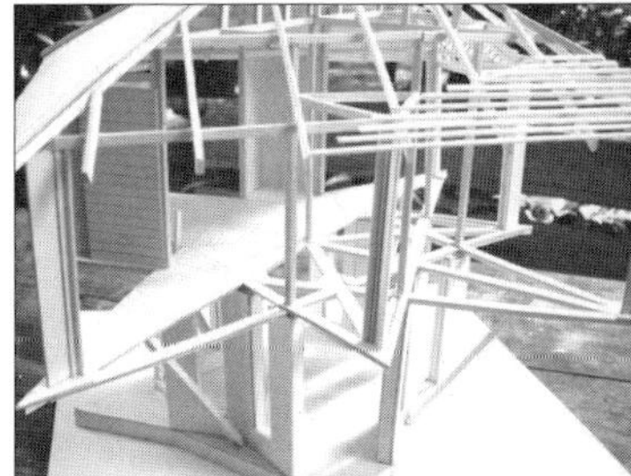

contractor and building supervisor Sethdawut Pinyorith was, again, a sheer pleasure.

Back in Stockholm I couldn't wait for another chance to visit The Land again, since I hadn't seen the completed house yet. It took another two years. In January 2007 I went to Chiangmai again. This time I went with my wife Marietta (with whom I initially had discussions about co-designing the house) and my four year old son Julius who had been playing around with the model of the house earlier in our home. The Land Foundation had also invited my fellow colleagues in the sound art project freq_out, an individual/collaborative project investigating the properties of frequency ranges within a larger spectrum in a specific space. The space this time was The Land. So suddenly, moving around amongst buffalo, rice crops and recently emerged buildings and constructions making recordings of animals, insects and reptiles, nature and constructions, were the likes of Benny Nilsen, Jana Winderen, Jim Thirlwell, Finnbogi and Tina Petursson, Kent Tankred, Jacob Kirkegaard, Brandon LaBelle and the whole family of Maia Urstad. Nico Dockx, Peter Verwimp and Kris Delacourt from Building Transmissions in Antwerp were also there merging their project with ours. Rirkrit and Kamin and the entire Land Foundation community prepared an outdoor evening dinner. Julius was playing around with architect Aroon Puritat and Marietta was talking to the frogs while munching on a piece of barbecued chicken. Sheer joy! I guess the only one missing was Thomas but we had just seen him in Bangkok getting married so he was excused.

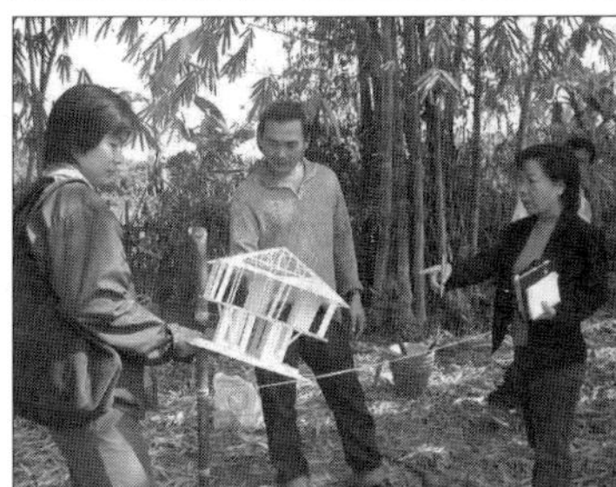

A few days later I went back to The Land with Julius, Franz Pomassl, photographer Anna Ceeh and Aom, the main vein at the Land Foundation office. I brought the green hammock that Marietta bought at Koh Chang in 1998 while her sons Georg and Erwin and I were snorkeling in the Bay of Siam. Julius and I set up the hammock on the first floor in the Starhouse and everything was perfect. The body swinging slowly as the smooth wind floated through the space while the bamboo trees gently swayed outside the house. Only a few details needed to be settled now: a bookshelf, a small solar power energy device and an MW radio receiver. The rest is up to the people who are supposed to hang around there and live.

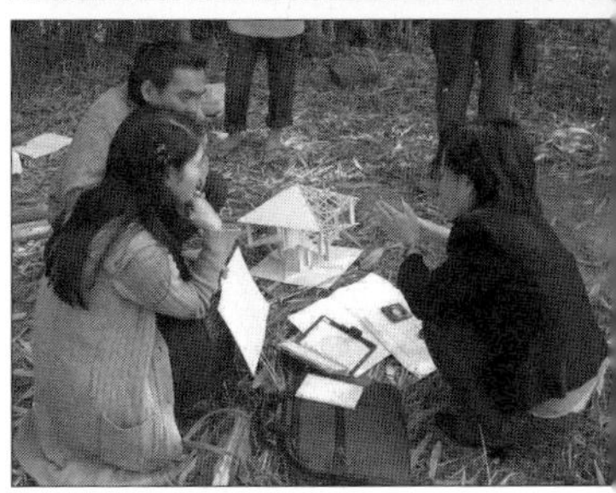

So what is this structure? I don't know really. I had it built but it's not mine. It belongs to something else. It is a part of the statement I mentioned above. A statement that doesn't present anything else but a strong will to live another type of life. A life without the norms of Western exploitation and competition, without pure hate and conquering and without loneliness and bitterness. Perhaps it's a proposal for a

33

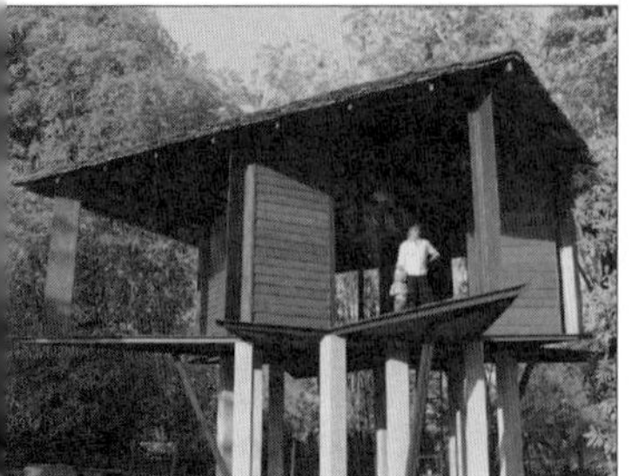

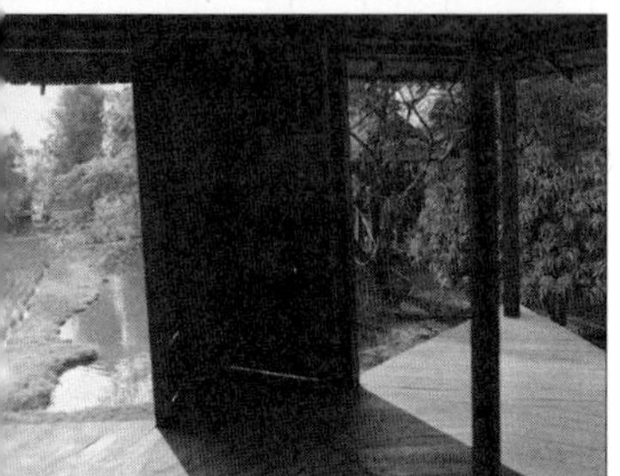

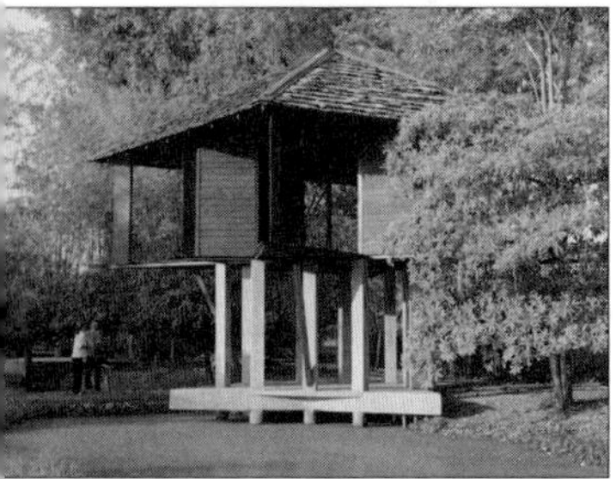

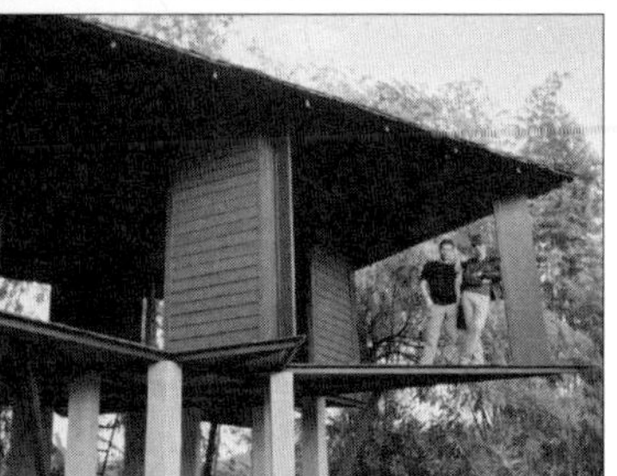

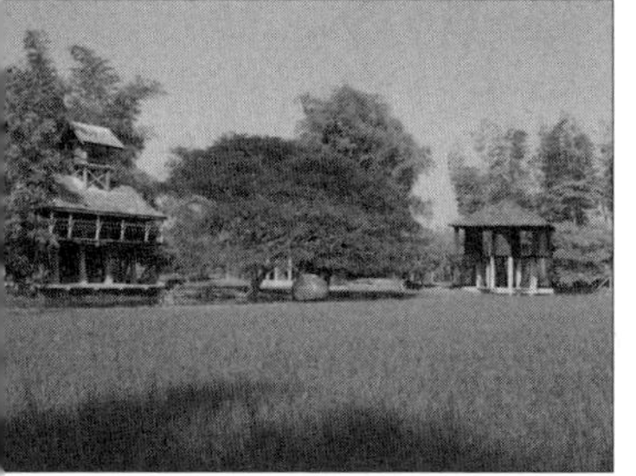

huge global society to be a bit more cool, a bit more generous and a bit more aware of not only themselves but also their surroundings. To try to stop being so crushed and destroyed and battered and helpless and low. As architect François Roche said: The Land is an icon. I guess it is an icon to take seriously then!

NOTES ON THE LAND FROM THE LAND

Initiated in 1998, the Land (a more direct translation from Thai to English would be "the rice field") was the result of the merging of ideas by different artists to cultivate a place of and for social engagement. The Land is located in proximity to the village of Sanpatong, a twenty-minute drive from the center of the provincial capital Chiang Mai, Thailand. As some rice farmers are having difficult times in the area, due to the levels of floods and high water level, rice farming has not been very productive. Because of this, some rice fields in the area have been offered for development, as the rice farmers are seeking to find better areas for the fields. Though initially the action to aquire the rice fields was initiated by two artists from Thailand, the Land has been developed with anonymity and without the concept of ownership. The Land was to be cultivated as an open space, though with certain intentions towards community, towards discussions and towards experimentation in other fields of thought.

The Land and its topographical environment (landscape) as it now stands was cultivated through the philosophy and argicultural technique of a Thai farmer by the name of Chaloui Kaewkong. The ideas around the cultivation of the topography, which is 1/4 earth (mass) and 3/4 water (liquid), are based on the composition of the human body. As there is much nearby water, an agricultural irrigation stream on the one side and a natural stream on the other, a series of ponds and pools were laid out in relation to the usage of water. In the middle of the Land lie two working rice fields, which maintain an ongoing relationship to the initial rice field itself. Rice has been grown and harvested yearly, though initiated as an experimental project to grow and harvest rice year round (rather than seasonally), and the harvest which yields in approximation 1000 to 1500 kilograms is shared by all participants involved and with some families in the local village that have fallen ill from AIDS. Presently these two fields are the most active part of the land. Fruits, trees and edible plants are dispersed into the landscape as well. Vegetables and different assortments of salad greens and herbs are also regularly planted.

There is no electricty or water, as it would be problematic in terms of land development in the area. As it is not the intention to develop the land for any value intrinsic to land development per se, the lack of such amenities was a simple solution. There will be development and experiments in using natural, renewable resources as sources for electricity and gas. The artist group Superflex from Copenhagen has been developing their idea of Supergas (a system utilising biomass, such as shit, to produce gas), and they have been using the Land as a lab for the development of their biogas system. The gas produced will initially be used for the stoves in the kitchen, as well as lamps for light. There is also an interest in developing a system for utilizing solar power, as another source of energy to be stored. These projects will engage interested participation from the local village as well as students from local schools and universities. Water is not a problem, however chemical pesticides and other such products have been introduced into the rice fields, which in turn feed into the water streams and system. In the center of the Land is an isolated pool of ground water, made from natural filtering of the ground content, however tests will have to be made throughout the year for any contaminations. Various pools and ponds will also be used for the farming of fish, a project to be initiated by a young Thai artist Prachya Phintong.

PRINCIPLES

SELF-SUSTAINABILITY

The freedom to express one's thoughts is one of the most important things in life. Freedom will be complete only if fundamental needs, such as food, clothing, shelter and medical needs, are fulfilled. In the lifestyle of the present society people rely on external economies, consumerism and competition to foster personal benefit, e.g. agriculture that uses industrial products such as chemicals and gene manipulation, without any realization of their effects on the environment and ecological systems. People only want to acquire more money and more physical convenience and comfort. This keeps people from looking within and achieving freedom because we need to rely on external factors and systems. The above observations are at the core of The Land Foundation's investigation and experimentation into adequate living and related systems, leading to attempts in self-sustainability as much as possible. It also promotes natural farming practices to revive a natural way of living.

SELF-CONTROL

Sometimes when we have too much freedom to express something we tend to forget ourselves and neglect other people's freedom. A perfect society must involve sharing and helping each other and respecting personal rights. These aspects will not occur if in the society each person does not understand real value within. Vipassana techniques of meditation enable us to understand causes and factors in being and value within. Once we respect and understand value within ourselves, it will bring about submission, acceptance of different opinions and therefore respect for other people's values.

DIALOGUE

Creativity is the heart of social development. To achieve perfect creativity the freedom to express and realize values in life related to nature are necessary. Creative communication originates from attentive listening, exchanging of information and respecting different opinions. These always empower the learning process. Art and culture are communication tools and lend to social creativity, which has made social development possible from the past to the present and will make it possible from the present to the future.

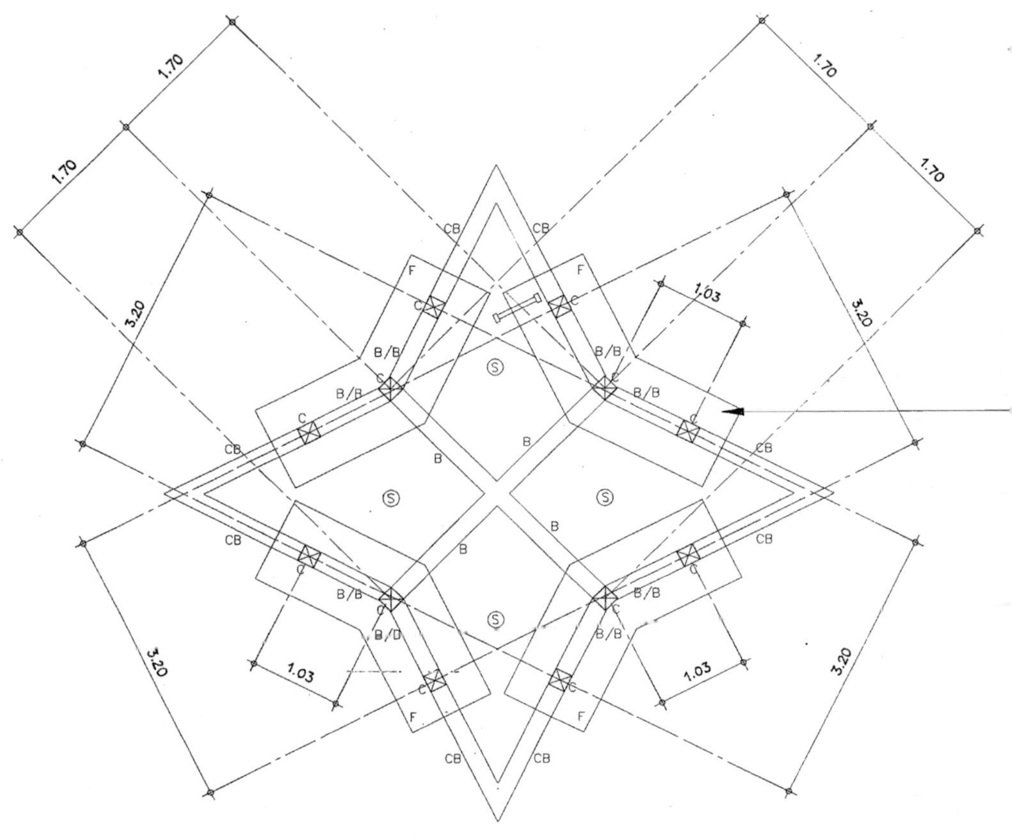

1.70
1.70
1.70
1.70
3.20
3.20
3.20
3.20
1.03
1.03
1.03
CB
CB
CB
CB
CB
CB
CB
CB
F
F
F
F
C
C
C
C
C
C
C
C
C
C
C
C
B/B
B/B
B/B
B/B
B/B
B/B
B/B
B/B
B
B
B
B
S
S
S
S

AROON PURITAT

One night in early May while acclimatizing to my first visit to New York I received an e-mail from Rirkrit Tiravanija. He was at that time in LA and he said that Michael (Carl Michael von Hausswolff) would like to have an article about the Land and his house there. Perhaps I could then add some ideas about the Land. After reading an attachment from Michael, the first image that flashed in my thoughts was that of Julius, Michael's dear son. That little naughty boy hugged tightly his Ultra-man doll, when he was in a lodging in the middle of a forest in Chiangmai. There was no Ultra-man in Sweden but Julius had Ultra-man printed on his pajamas. He had a few more Ultra-man dolls laid among other super-heroes from elsewhere in the world in his personal toy-suitcase. Julius was absorbed by a VCD playing animation of Ultra-man, though the language uttered was clearly Thai, at the moment the hero let go enormous light energy against a monster. My thought wandered to an argument on the issue of intellectual copyright between Thailand and Japan. It was claimed that Ultra-man was created with inspiration from a Sukothai Buddha image. It seemed impossible to me. The visit of Michael's whole family was like that of the journeys of families or the moves of foreigners to Thailand which are familiar sights and have become so common. However looking at this little blonde boy in Ultra-man pajamas led me to think of numbers of stories when the Land was first established. That was also the beginning of the movement of people flying from other parts of the globe to Chiangmai.

The image of the Land is continuously spread via 'media' all around the world as a model architectural structure, an ideal space and a utopian community. However, the individual aspect is left out because people might not strongly believe in the notion of an ideal person. It seems impossible to believe that someone who is still alive could act as a prophet, as a figure who would bring permanent peace into this world. Taking Buddhism into consideration, this 'world' travels gradually toward decadence, every moment the physical world deteriorates and challenges our ability to accept the natural order of decline. People involved in the Land usually are interested in certain contents while the level of interest differs with each individual. However, the most important thing, which should be accepted, is that people involved with the Land have the ability to achieve a high level of 'media' output and can choose which 'media' to use. Thus, when media refers to the Land, the dimension of sociality and the unavoidable reality of family institutions might be disregarded, for this dimension supposes that a natural mechanism enables human existence to develop later into forms of community. In contrast, the gathering of people through technological networking may portray an image close to an 'ideological community' like the Land, where diverse individuals meet and gather through informal contact to create a platform for sharing. But that doesn't mean that each individual attending this space will slip off the structure of family institutions, because family institutions take into account sensitive mental complexities and bond them to other problems, forging a chain without end. Unless the individual tries to understand his or her own family institution and create certain conditions agreeable to family members so as to keep a kind of balance, the familial bonds will have to be taken away from one's mind in order to further carry on his/her journey to the ideal territory. And another condition is to let the mechanism of deterioration play its role and eventually release oneself out of this structure.

NEW YORK-BANGKOK-CHIANGMAI
MAY–JULY 2007

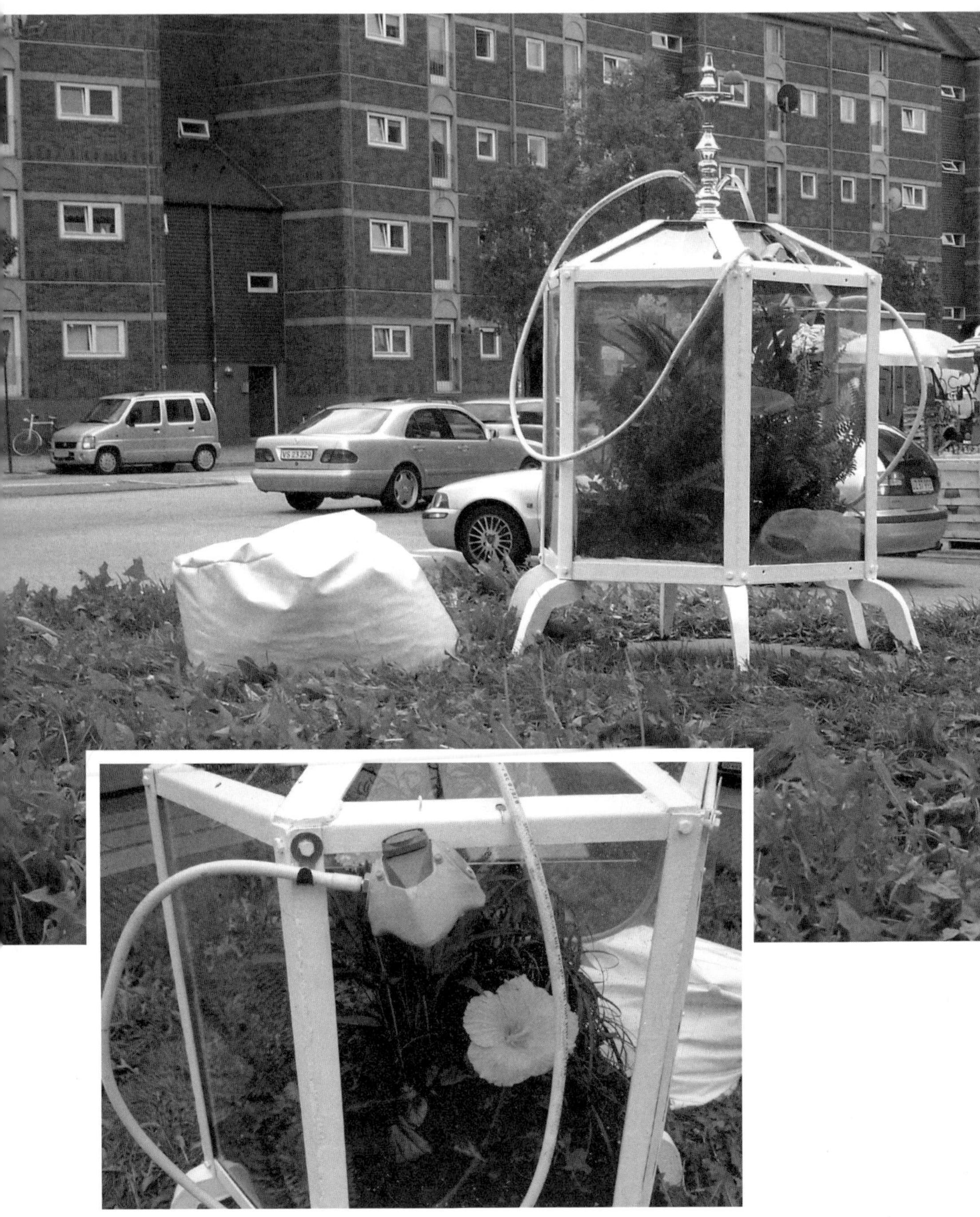

Hartmut Stockter: *Oxygen Greenhouse*

Nis Rømer

Hot Summer of Urban Farming

The "five finger plan" of Copenhagen envisions the city developing along five lines of infrastructure and leaves intact green belts between the fingers deep into the city. This quite ingenious plan holds the promise that the city and its surroundings can be integrated and the growth of the city controlled to some degree by planning the infrastructure. Developed in 1947, the plan still works as the overall framework for the city, though much has been built between the fingers.

The *Hot Summer of Urban Farming* project continued along this trajectory. As curator of the project I was interested to explore ways that gardens and farming could be integrated in the city, though on a much smaller scale. Over the summer of 2006 eight artists and artist groups spent time making plantings and temporary gardens in Copenhagen's lesser-used spaces. The artists dealt with issues of inclusion and exclusion, the dynamics of public space, the origin and history of plant life and the relation between the city and its surroundings.

The aim of *Hot Summer* was not to actualize all the potential for urban gardening; the aim was rather to point to it and discuss urban spatial politics by way of example. The brief for the artists was to make examples through small plantings or other interventions that worked as images, and then to make a poster that took the ideas further or exemplified the related research. The posters were silkscreened and distributed in the city as a way of inserting ideas into the public sphere. In this way the project re-imagined and created images that expanded the discussion of what a city can look like, how it can be interactive, and how we can imagine public spaces. I think in general this is a potential in art that is often underused: art and politics are in a way parallel endeavors in imagining and defining what is possible and thinkable in contemporary society.

The works were situated in an area of Copenhagen, Outer Nørrebro, which is characterized by the highest density of people in the city, along with the largest ethnic mix and the least square meter of public space per inhabitant. But if you take a careful look there are small pockets of indeterminate spaces mainly used by dog owners, if at all. In the tradition of "zwischennutzung" or temporary uses of city spaces, the projects the artists made were temporary or made to slowly dissolve into the city and develop a life of their own.

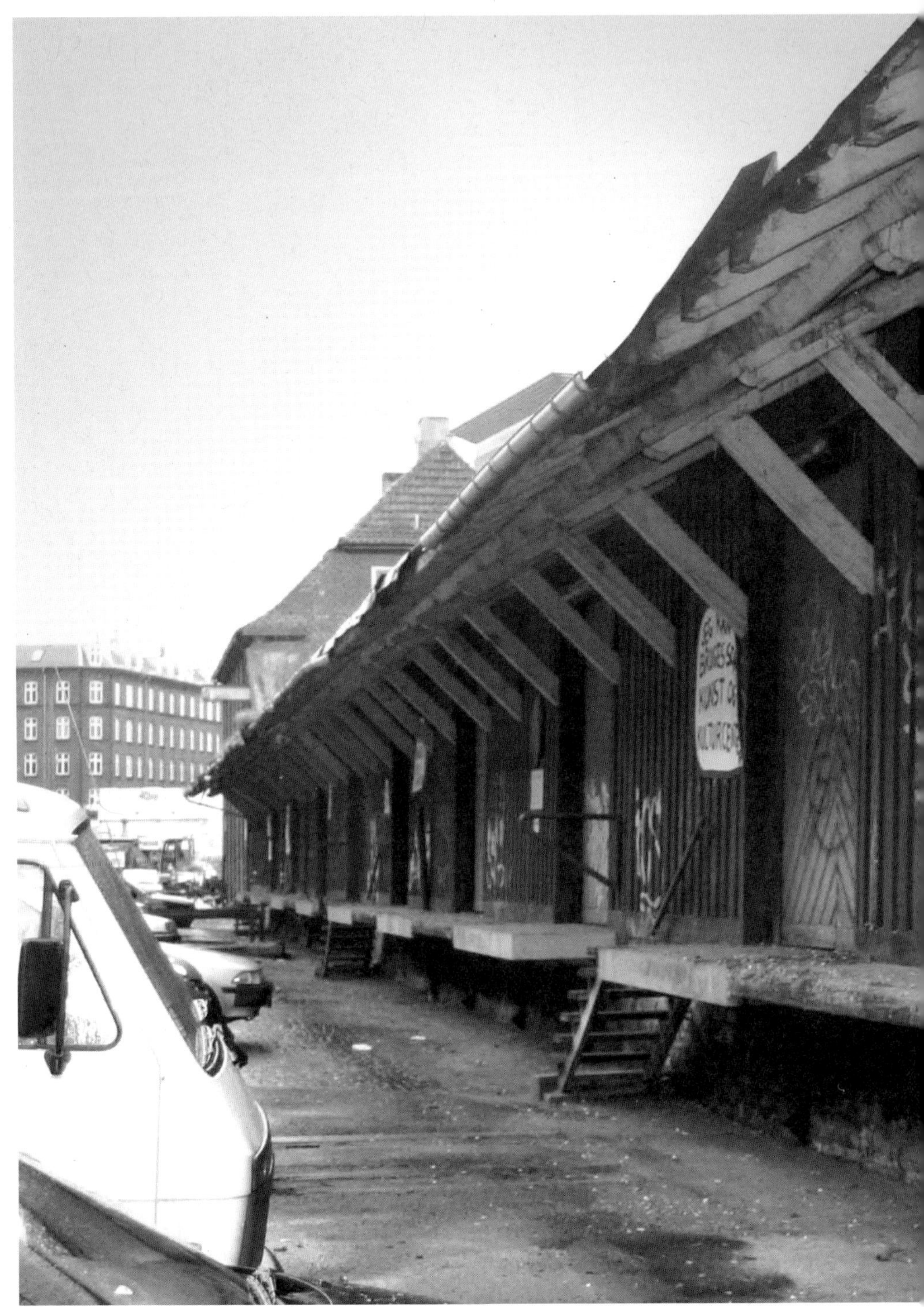

YNKB: A cultural center at the Freightman Halls

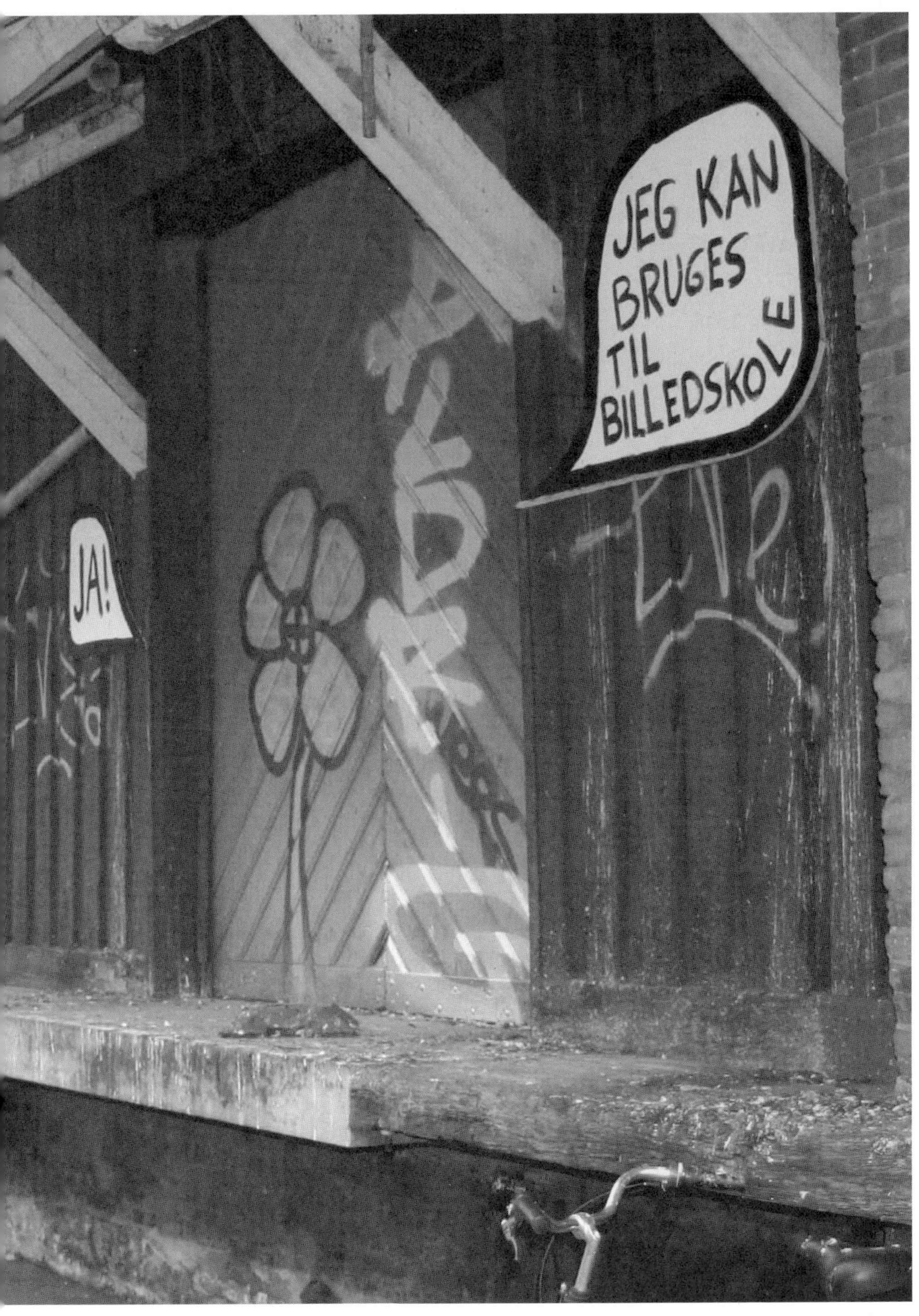

NIS RØMER

Space and the use of it within the core of the city have often been contested. In Copenhagen, "Ungdomshuset" or Youth house, a place for self-organized music, culture and politics was torn down in March 2007 by a massive police mobilization, which was followed by demonstrations and street fighting, especially in Nørrebro. The Youth House was one of the very few places in the city to have a public, self-organized garden with herbs, flowers and a fireplace. For many years this was one of the only community gardens in Copenhagen, and the house carried an important history of self-organization. In a country like Denmark where the (welfare) state is so strong, this is quite an odd concept to most, as the general understanding is that everything public is also something that the city or the state takes care of. In the same way, the concept of being a co-creator of space through gardening or making public spectacles is generally unthinkable.

The demolition of the Youth House made it apparent to most that the use of and access to city spaces is a highly political issue, and that the emphasis on property rights over all other rights favors normalization and eviction of groups that think, look and behave differently. The term "normalization" is even used by the government for the process of changing another of the city's great areas of difference: Christiania, a military base squatted in 1972 that has continued until today as an experiment in alternative housing, urbanism and local democracy. While it seems a global trend that the pressure on the ground in big cities is increasing, it is not a given that this needs to happen at the expense of supporting differences and their expression. But it takes not only strong local organizations around the spaces that exist, and ideas for how they could be used, but also awareness on the part of the politicians of the value of local green spaces.

Spatial politics and spatial justice have become common terms in art and grass roots organizations alike. This is a level of politics that has a direct effect on people's quality of life in cities. Even a few square meters of free space in a neighborhood can have a life-changing effect for children and adults; these can be spaces that are not regulated by institutions or commerce and dedicated to what most people experience as the most important part of their lives: social relations, free expression and playing and learning from others. Spatial justice concerns the fact that access to these vital resources should be equally distributed and not limited to the wealthy areas of the city.

A project that touches on the distribution of space is Jonas Maria Schul's *2200 N Parterre*, which overlays a garden landscape from a historical royal castle outside Copenhagen on an empty lot used by dog owners to run their dogs and that generally appears as an urban litter garden in a traditionally low-income part of the city. A "parterre" is a classical French gardening style where royal monograms or other patterns are created through flowers and low bushes. By rolling out ready-made grass mats a monogram was created, blended into the Nørrebro plot, but also conditioned by the neighboring weeds. Jonas explains that the garden was made "to create a short odd moment; a clash between cultural spheres, but maybe also a silent, distressed protest against a development where immense gaps between rich and poor are shaped by the political elite."

HOT SUMMER OF SUMMER FARMING

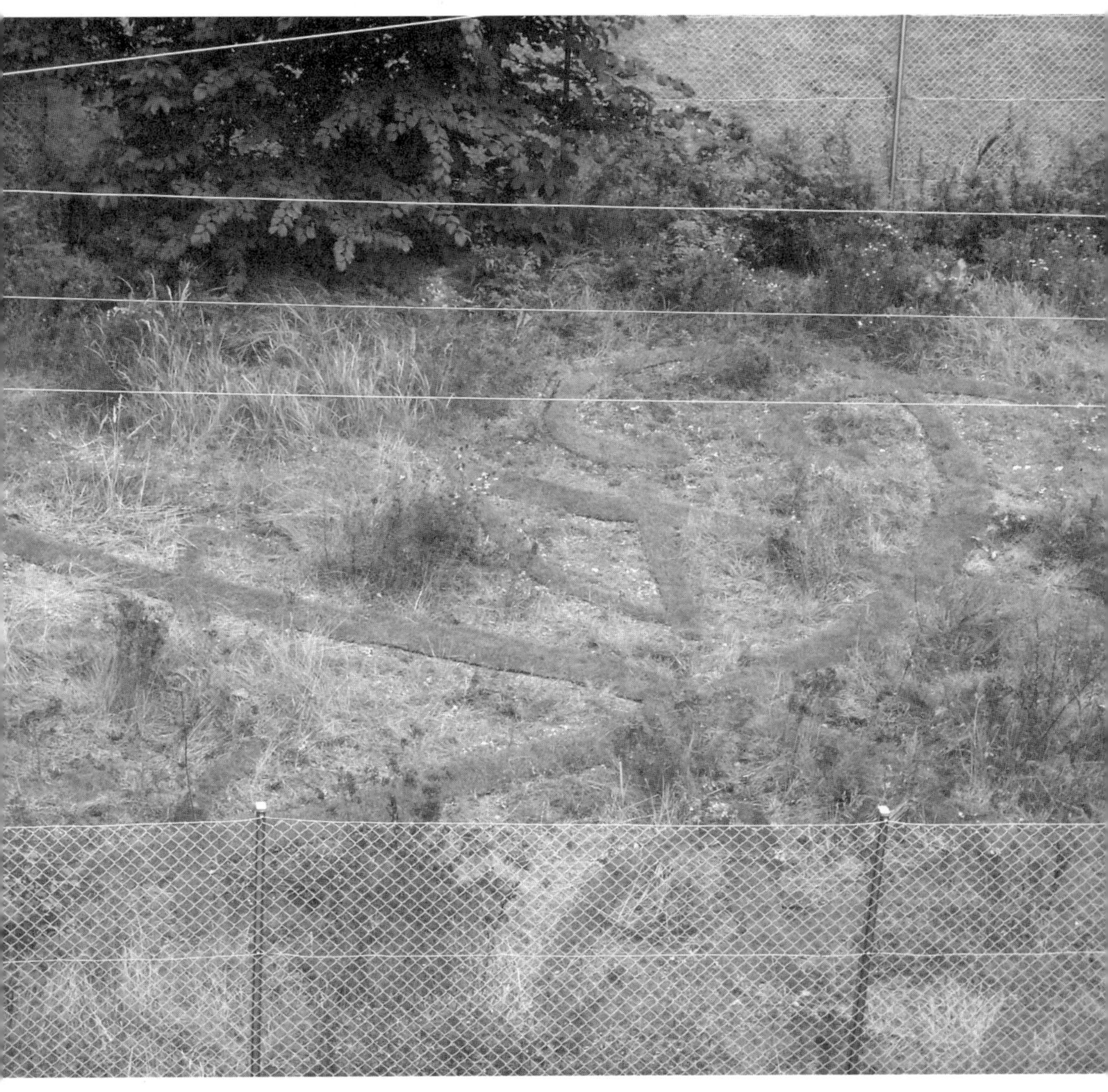

Jonas Maria Schull: *2200N Parterre*

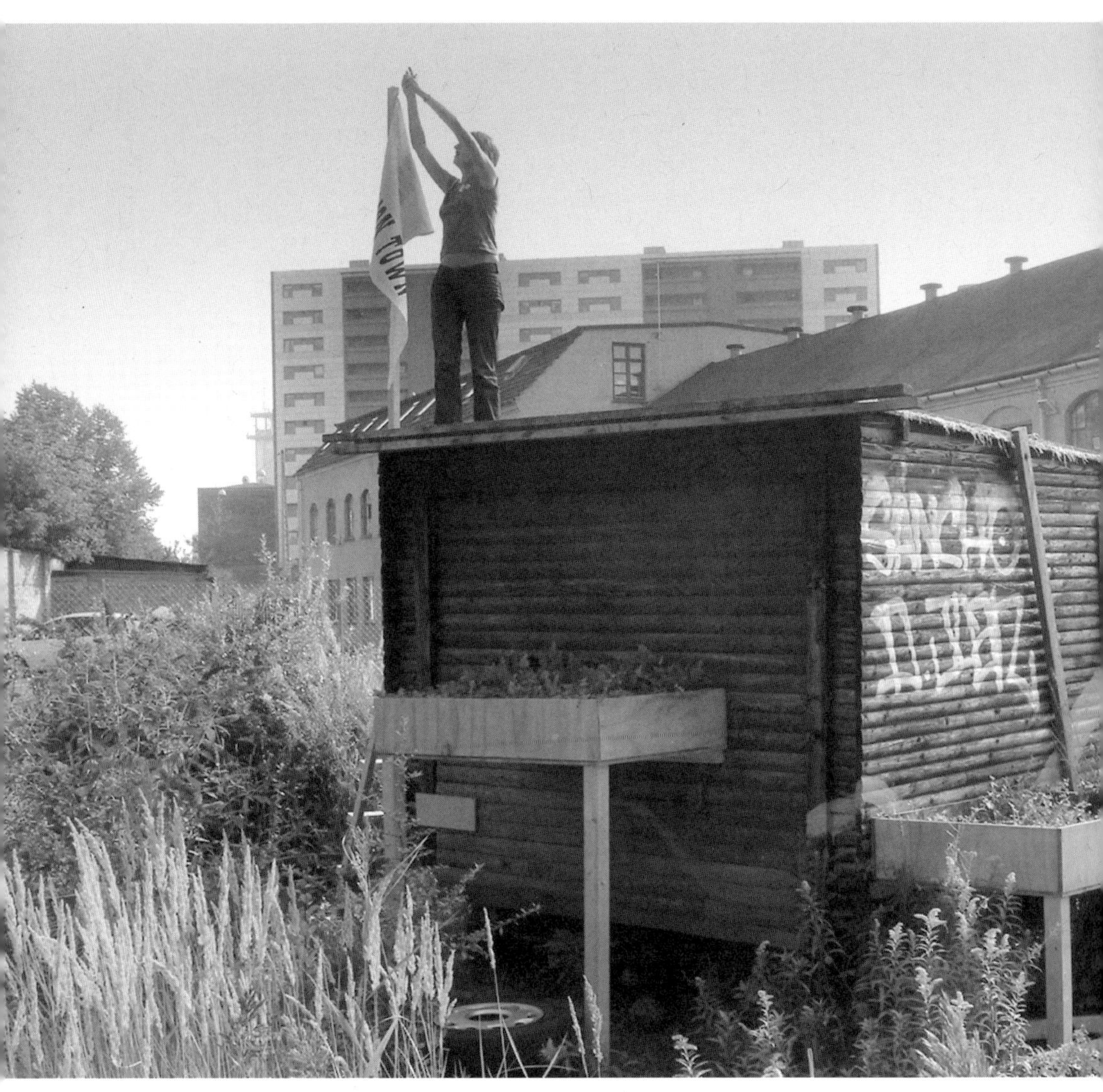

Camilla Berner: *Dandelion Town*

The Artist Group YNKB have documented on the *Hot Summer* website their year-long effort to make a cultural center and activity park on a former railroad area and freight hall centrally located in Nørrebro. This area is one of the biggest unused spaces in the city, and through a whole array of legal actions, meetings with politicians, local organization and rallies, they have managed to keep the issue present in the minds of the public and local politicians alike. The mix of activism with art and culture, screening, interventions and exhibitions as well as alternative planning efforts makes this quite a

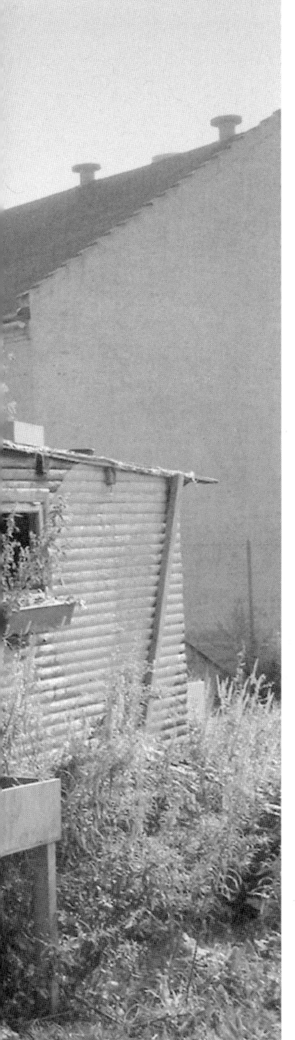

unique project. At one point they actually succeeded in having the building declared a historical landmark, which meant that it should be maintained and restored. But the now privatized railroad company managed to overturn the decision and as a result the old wooden halls have been torn down.

Sustainability is likewise an important issue in discussing urban farming. Ideally, food grown in the city would reduce CO_2 emissions, as the food need not be transported from the place of growth to the place of consumption. Artist Marie Markman, as part of her contribution for *Hot Summer of Urban Farming*, calculated how much could actually be grown if the leftover spaces of Outer Nørrebro were cultivated. While the project does not suggest that this should actually happen, it would provide for some weird parks and plantings in the city if other crops were introduced. The idea of the edible park comes to mind and would for sure look quite different from most inner city parks made for easy maintenance and leisure only. Through walking and measuring leftover and in between areas on maps, Markman calculated that there are 40,000m2 of un-utilized space in Outer Nørrebro, approximately 1m2 per inhabitant in the area. Calculating further, she found that this could yield 80 tons of potatoes each year or supply all kindergartens in the area with carrots.

A different approach was taken by Camilla Berner with her *Dandelion Town*, which suggested a whole city taken over by dandelions. On a location that has been unused for as long as anyone from the area remembered, she expanded on an unused shed with experimental plantings using the often-undesired plant. By depriving the plant of light you get an almost white plant delicate enough to use in salads. She made dandelion marmalade and over the time of the exhibition built up a massive online knowledge bank with recipes and experiences from the process. The poster for her project describes a diverse city where high rises, housing and sheds are connected by ropes and ladders, and all are bound together by dandelions.

The artist Åsa Sonjasdotter made potato gardens with a variety supplied by the Nordic Gene Bank, a remarkable institution that stores seeds and plants from all over the world. Based on thorough research into the cultural history of the potato, the work tells a story of how patents to plants are managing to make it illegal to grow most varieties of potatoes. One sort has a special status: rights for the "Asparagus Potato" have been bought by an independent group and given back to the public. The history of the potato was traced through the encyclopedia of Diderot who claimed that, "However it is prepared this root is tasteless and starchy. One could not include it among the agreeable foods."

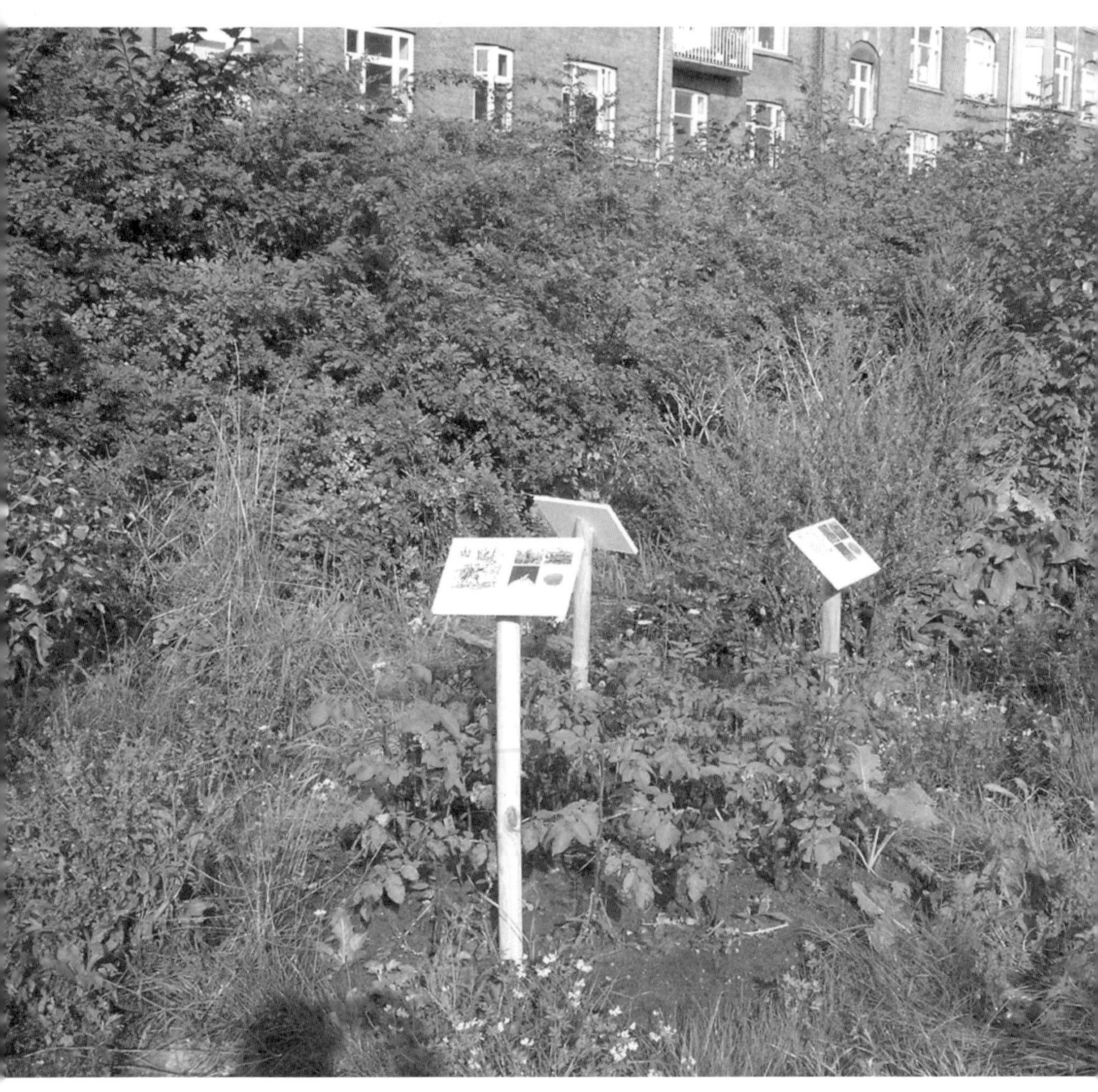

Åsa Sonjasdotter: *Potato Perspective*

Finally, Sonjasdotter made a series of gardens and grew varieties seldom seen, like a blue potato, and accompanied them with signs about the cultural history and legal status of the different potatoes.

Nance Klehm from Chicago made a project about immigration and farming; she exported corn from the US into Denmark, made it into tortillas and served them in exchange for stories of migration. She recreated the whole legal process of applying to the US ministry of agriculture to get permission to

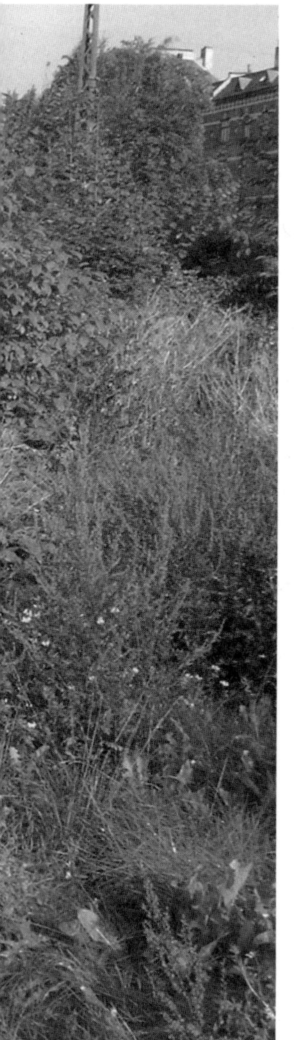

send a few kilos of corn to Denmark, which mirrored the legal difficulties that follow immigrants when traveling across borders. In Copenhagen she made a mobile kitchen that could be attached to a bicycle and in the process met people with whom she exchanged services and stories, so performing a temporary integration into Danish society. The stories she got in exchange for serving the handmade tortillas in the city can be heard on the website.

Gillion Grantsaan took the approach of an active user of public spaces and a collector of popular stories and memories often shaped by political events from afar. As the population of Outer Nørrebro has a high percentage of immigrants, international politics, war and unrest are an important part of everyday life here often heard through media, but also from relatives that live in troubled zones in the Middle East and beyond. What to Danish citizens is often just background media noise takes on a presence and a voice. Gillion listened and recorded stories and made them into a website documenting parallel realities.

The Oxygen Greenhouse made by Hartmut Stockter was the stationary version of the Plant backpack (a portable greenhouse rucksack attached to a breathing mask), which was designed for the urban day-tripper yearning for fresh air. The daytripper is the alter ego of the artist engaged with science and inventions and how they can be applied for everyday uses. The greenhouse was designed as a mix of the classical greenhouse and oriental architecture and referred to the shawarma bars and the dome-like architecture of the neighboring Nørrebro Station. A tongue-in-cheek comment about how we organize the city and a longing for a purity that might be long gone, at the same time it offered exciting new mixes and possibilities.

Hot Summer of Urban Farming was also thought of as an exhibition that gathered people with an interest in the environment within cities and in what the shaping, cultivation and sometimes destruction of the urban landscape has to tell us about contemporary politics, culture and history. The relation to the landscape is one that defines culture and indeed is one of the most important and disputed relationships in our society. This differs from much previous thinking on the landscape in early Land Art where nature and culture were perceived as separate spheres and where Land Art was more often than not something that took place way outside cities. The shift to working in cities and considering the extended landscape marks a change in thinking away from a notion of site-specificity and toward a more complex notion of urban environments and free public art.

All projects can be seen on the website: www.publik.dk/hotsummer

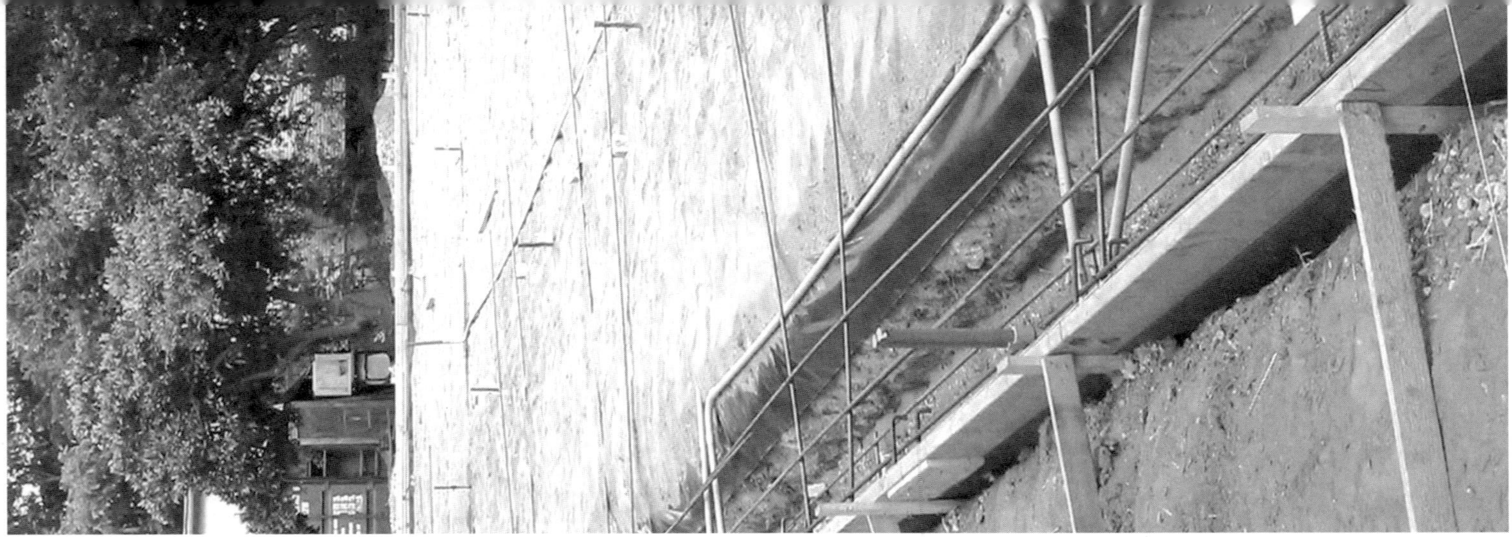

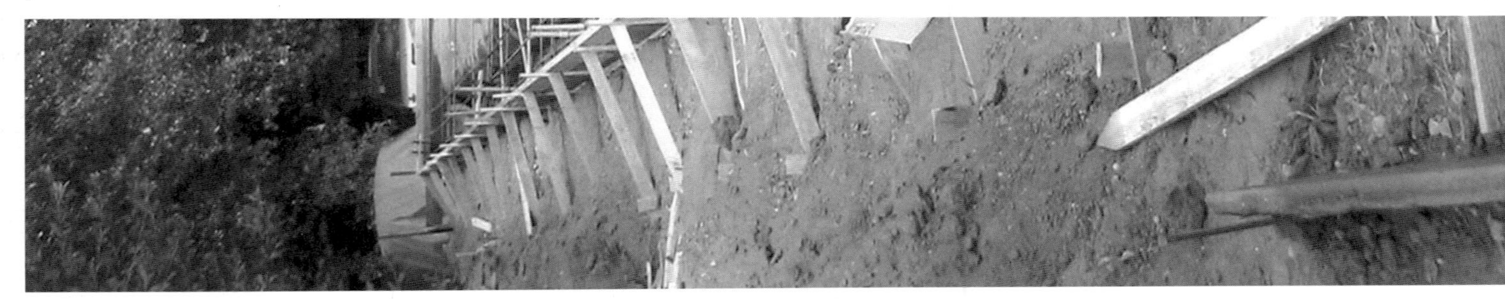

The site has been graded, and formwork and reinforcement laid out. This is just before the concrete slab foundation was poured. Looks kind of like the stretching of a canvas if you ask me. Something about the moment before, or maybe just the need to pretend there's a before, before one can start.

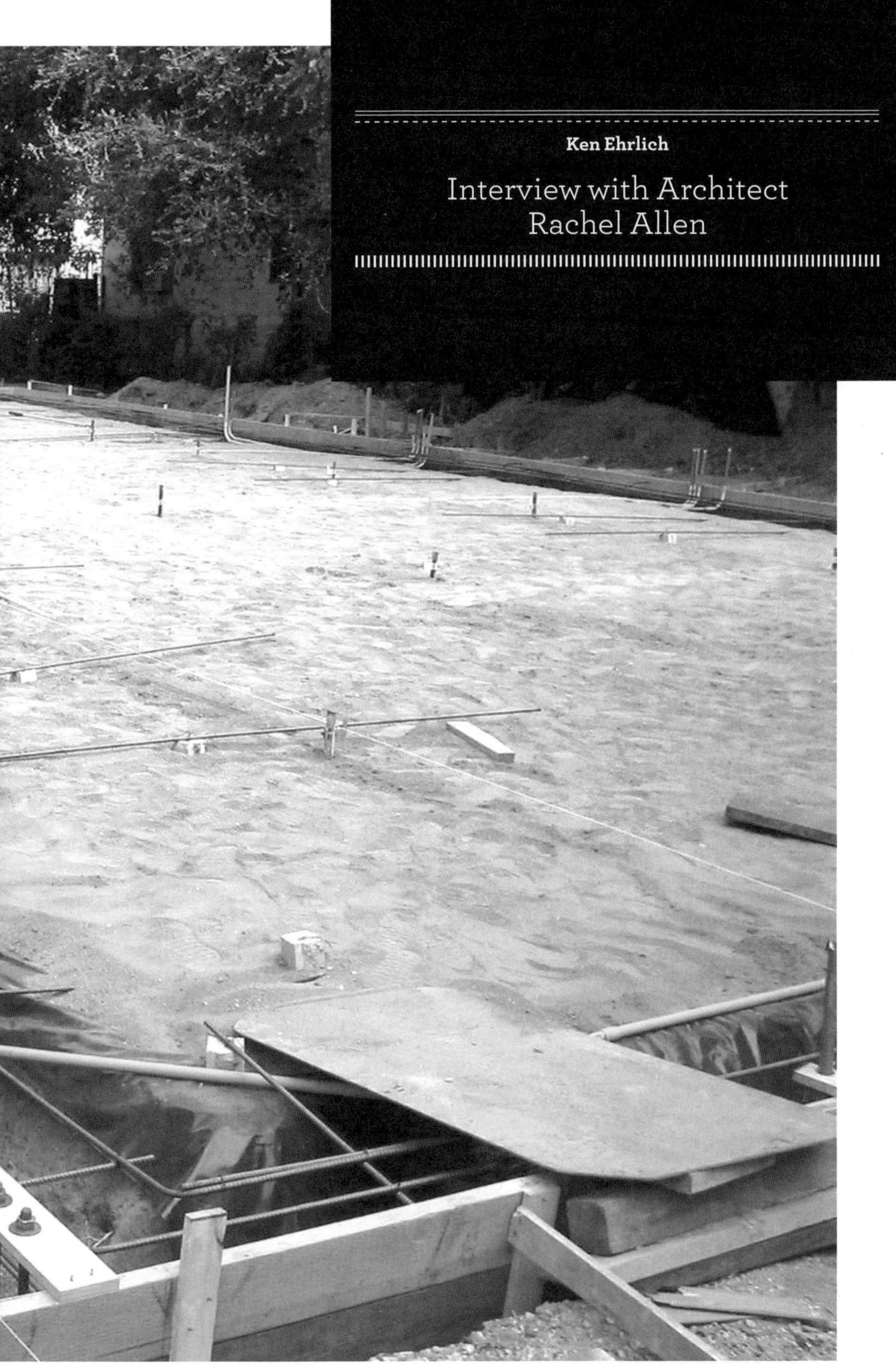

Interview with Architect Rachel Allen

I first met Rachel Allen in 2004 during the *October Surprise: Creative Interventions and Underground Politics* in Northeast Los Angeles, a series of installations and interventions in a rapidly changing part of the city I helped organize with a number of other artists and activists. Rachel participated in the event by offering her architectural services to the public for free in the form of short brainstorming and drawing sessions. Her contribution, entitled *Dream House*, connected individual spatial desires and fantasies to the larger context of *October Surprise*, which included issues such as land use conflicts, economic development and neighborhood histories. Since that time, I have followed her work with interest. Initially I requested that she submit a visual contribution for this publication that addressed the relationship between art and architecture. Questions about her contribution evolved into a substantive e-mail interview, an edited version of which is published here.

<u>KE:</u> The three images you've selected for publication offer views of a studio in progressive stages of completion. It seems that the perspective of the camera (and viewer) shifts around the space as it's being built. This made me think about the creation of a space and how that is related to temporal dynamics, the way a space changes over time. Both art and architecture face this issue of whether or not (or how to) reveal the structuring devices of a particular work. To what degree do we reveal the fabrication of the work as part of the work? I'm wondering if you might say a bit more about how this work operates for you conceptually: I am curious how you see it fitting into the larger context of art, architecture and the social field. How do the images speak to the dynamics between disciplines to you?

<u>RA:</u> I chose images of the studio we designed for the filmmaker and artist Pat O'Neill, called Lookout Mountain.

The best way we could think of to address the relationship between art and architecture was to document a place where we had put the two into an active, everyday process of relating. Rather than asserting anything or didactically claiming anything about them as disciplines, to start out by focusing on an instance of them in relationship. Like a case study.

Combing through our records, we found the first two pictures and started to think of other things that are common to the two projects of fine art making and constructing buildings.

These images don't deliberately speak to the dynamics between art and architecture as disciplines. That doesn't really interest me.

But I'll pause to ask you, should it? Am I missing something here? Why does it interest you?

After all this time I still like the Beaux-Arts model the best: The fine arts (music, dance and theater, painting and sculpture, poetry, architecture, however one likes to tally them, as this list used to include gem-cutting after all) are pretty sisters, all dancing together in a parade. I am a closet anti-interdisciplinarian.

I should also mention that for me these two disciplines in particular are clearly distinct and the boundary between them as projects and as identifications for individuals or tribes (artists and architects), is not confusing at all. The training is different, the skill-set is different, the agendas and problems and goals are different.

This has become a necessary refrain in response to the commonly-heard remark that "architects all just want to be artists." This is such a silly thing to say. Some architects may think they want to be, and they may even say so, but that desire and those statements are rhetorical and need to be approached as such. As for me, if I wanted to be an artist, I would be. In that old yarn the figure of the artist is being used as a straw man to represent self-expression, and the architect represents the bureaucracy and business which is a part of every discipline. These substitutions are trite and not of value to either tribe. In response I'm forced to parry that I often find architecture by artists a subset of folk architecture—charming, but, like folk art, essentially conservative. Of course there are exceptions.

<u>KE:</u> I must confess that I have a sometimes hasty dislike for conventions. So when I think of conventional architecture I always think of the British critic who said " Architecture is window-dressing for the accumulation of global capital." My interest in crossing

disciplinary boundaries is partly inspired by the search for modes of working that do not re-produce the systematically corrupt arrangements that seem endemic to our economic system. This is not exclusive to architecture as a discipline. In fact, the art world more and more is dominated by an interest in "cultured" investing. And without sounding too much like a vulgar marxist, this speculation tends to dictate what is meaningful in the realm of contemporary art. You don't have to look very far in *either* discipline to see examples of trite formalism and echoes of crass materialism. This is not to say that interdisciplinary work automatically avoids all of the pitfalls of trying to do meaningful work in a cultural context that is set up according to structures that seem to mirror neo-liberal politics (the logic that profit margins trump all else, the disappearance of public funding for cultural work, and the appropriation of cultural work by a set of networks where the highly abstract internationalization of finance determines what is celebrated and what is irrelevant). Of course, things are more complicated than all of that and there are great examples of work produced in a rather "conventional" way both in art and architecture that I find exciting and meaningful. The Koolhaas library in Seattle is one example of a work that manages to operate within the problematic terrain that I outline above while working quite well as public architecture. I agree that in some sense there is no confusion between the disciplines. As you say, the training, skill-sets, problems and goals are completely different. However some of the work that I find inspiring explicitly challenges the line between these fields... I'm thinking of people like Buckminster Fuller

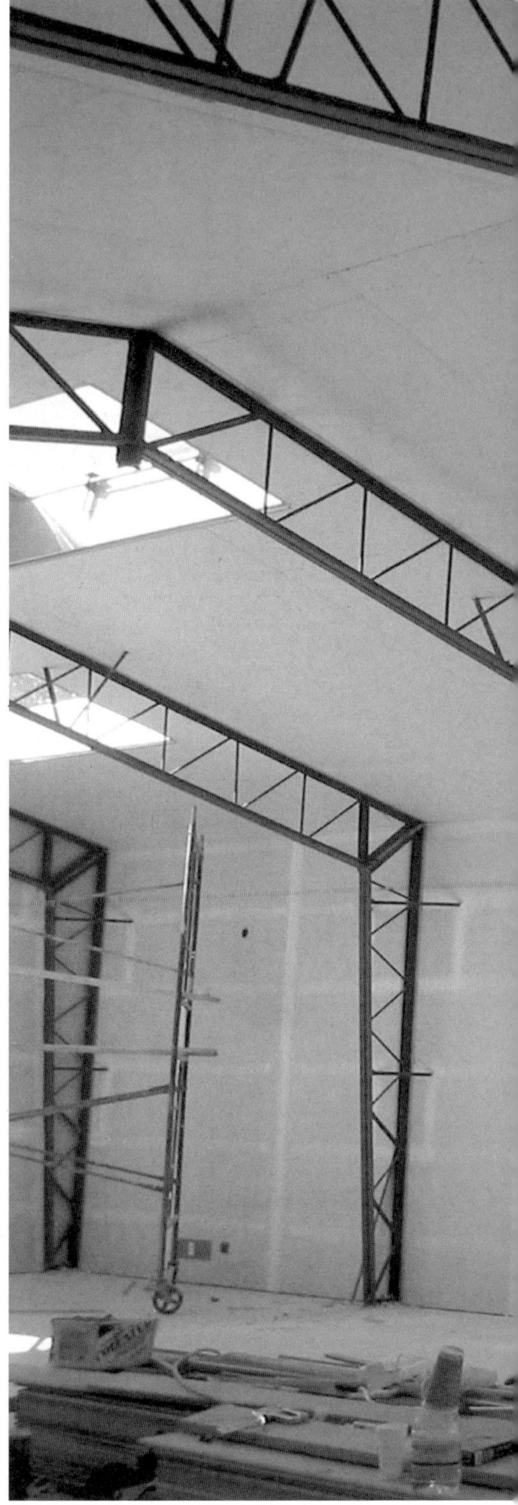

INTERVIEW WITH ARCHITECT
RACHEL ALLEN

The interior walls just before the paint goes on. Drywall's been taped and prepped. Again, paint is something that's common, and preparation, and the concealment of the process of assembly and parts, as you pointed out.

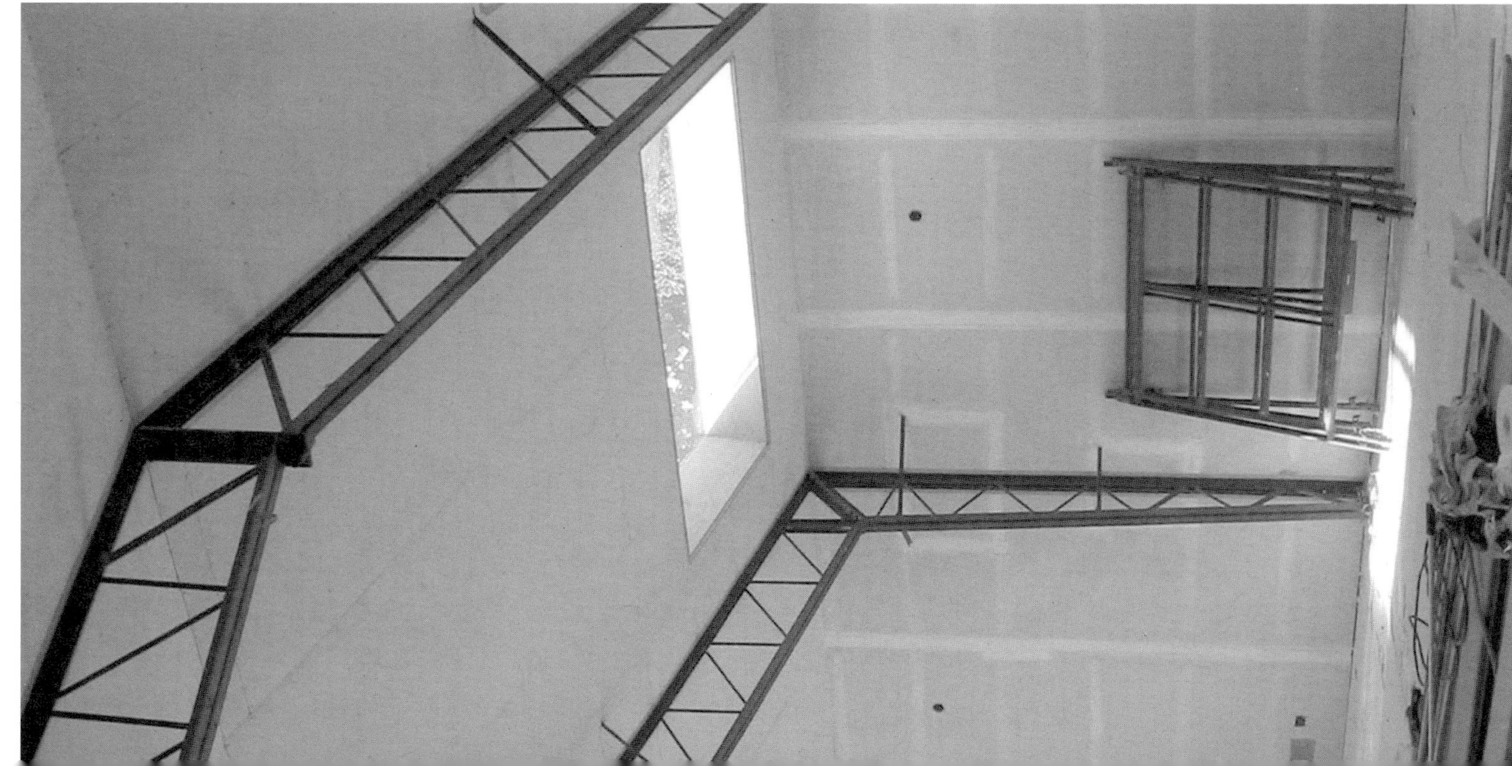

or Gordon Matta-Clark, two obvious examples. You begin to see that I'm somewhat of an idealist: A good architect can design an elegant building… but so what? A successful artist makes work that many people find beautiful… but is that enough? I guess if I'm forced to really explain my interest in interdisciplinary work it's based on a desire both to see and make work that challenges the disciplinary and spatial conventions of spectacular culture.

RA: I couldn't agree with you more about the need to work in ways that are not corrupt or stale or oppressive. But it has honestly never occurred to me that interdisciplinarity might be a way to avoid this. If anything, moving or crossing between fields has always seemed to me to dull my instrument, not sharpen it. And what could be more conventional than that these days?

Instead I value those conventions that seem to me valuable and ignore those that seem to me not so.

Perhaps I simply have a different relation to convention than you. Is that what we are uncovering here? I don't have that knee-jerk response, at least not where work is concerned. (Life is another story. Related, of course, but different). Depending on the convention. I have found that once I have a convention mastered (yes, I mean "mastered"), a drawing convention, for example, I find it so empowering. I love AutoCAD. I love spreadsheets, and agendas, and meeting minutes. When I move into another field, my inexperience often feels like a liability.

Elegance and beauty have never been the only end goals available within convention. I like it best when a work takes off into that weird outer space orbit around/outside the dualism of formal/beautiful and challenging/meaningful, such that I can no longer tell which aspect of the thing is performing which of these roles. (An incomplete list but some artists whose work operates this way for me: Caravaggio, Palladio, Yoko Ono, Neil Young, Ellsworth Kelly, Eugene Atget, Marcel Breuer, Michael Snow, Aalvar Aalto, and Agnes Varda.) And it seems to me the coolest thing that different things do this for different people, and even that an artist's work can do it when that same artist as a person can't.

Your Koolhaas example is so apropos. He represents a return to convention, in so many ways, as an architect. He'll work for anyone, the Chinese government, in Dubai, the Prada stores, he has no qualms and makes no apologies. But his projects are poison-pen love letters to capital every time, bills his clients open up and pay.

KE: The other thing that I'm invested in that is relevant both to art and architecture is the notion of *reception*. Granted it's another very complicated term…but the question of how we experience space and how historical conventions shape our perceptions is not something that is typically taken up within the confines of artistic or architectural practice. From Duchamp through Cage and Fluxus and throughout Minimalism and Conceptualism, however, there is a recurring consideration of architectural space and the impact it has on viewing art. Perhaps successful interdisciplinary work might continue to approach the notion of reception in alternative ways.

RA: I thought about this a lot, and I must confess that the question of reception and how we experience space, how our experience is shaped, has for me been relegated to the realm of the unanswerable, at least by me. What "we", what "our", who is the subject in that always passive voiced "is shaped"? I'm interested in being that shaper, and I give up on understanding exactly how it works, I only know that it does because I can occasionally observe it in action, when I see the effects of the causes I set in motion. I take up that question by contributing new objects to the world that try to do that job in strange new ways, ways I've never seen before, ways that interest me. But I don't know how they work, I admit that freely, and I gave up long ago on controlling or determining the process.

I suppose here I am influenced by Benjamin's pointing out that architecture operates on the unconscious, which only makes it all the more thorough and dangerous as a medium.

KE: Two things I'd like to follow up on in our previous exchanges: One is whether what we've called interdisciplinary approaches as a strategy is a productive model or a useful way to re-think assumptions in a particular discipline. Can you say more about this idea of mastering your field and how that is important to you? Also, I'd like to pick up on the idea of reception. As you know, within the context of art, notions of participation, interactivity and relational aesthetics have a certain charge and I'm wondering if within the field of architecture there is a parallel kind of thinking going on. I think the Benjamin quote about architecture operating on the unconscious is a perfect starting point. That is precisely why ideas about reception are interesting... because they are ambiguous, loaded and verifiable only through experience. And lastly, I re-

cently heard Wolf Prix lecture at SCI-Arc and he talked about LeCorbusier as a sculptor in very reverent ways... again this idea that architecture yearns for the status of art, Curious. Do you have thoughts about this yearning? Why do architects look to sculpture as a model for designing buildings? Gordon Matta Clark is famously quoted as saying about the difference between architecture and sculpture is *plumbing*. What is the difference to you?

<u>**RA:**</u> **INTERDISCIPLINARITY**

My formative education was from a solidly and self-consciously interdisciplinist standpoint. I took the founding seminars of the European Cultural Studies program at Princeton and was deeply influenced. Nearly all the reading lists in the Architecture School there at that time started or ended in Foucault, and I attended a lot of the important conferences that brought interdisciplinarity to the center of the architecture academy ("Sexuality and Space," etc.). So probably I don't even know what non- or pre-interdisciplinism would look like, at some level.

But what I was trying to stress in that remark was that for me since leaving school it has been the process of mastering the technical aspects of my field (and I mean this in the broadest possible sense, from building construction engineering to drafting to writing meeting minutes—all things that there were not courses in, at least not at my school) that provides a pleasure more affirmative and reliable than any other I've tried. And pleasure is more and more now how I organize my studio practice: follow the pleasure. If it's rewarding, do it again, and more of it, invariably it enriches itself or takes an unexpected turn.

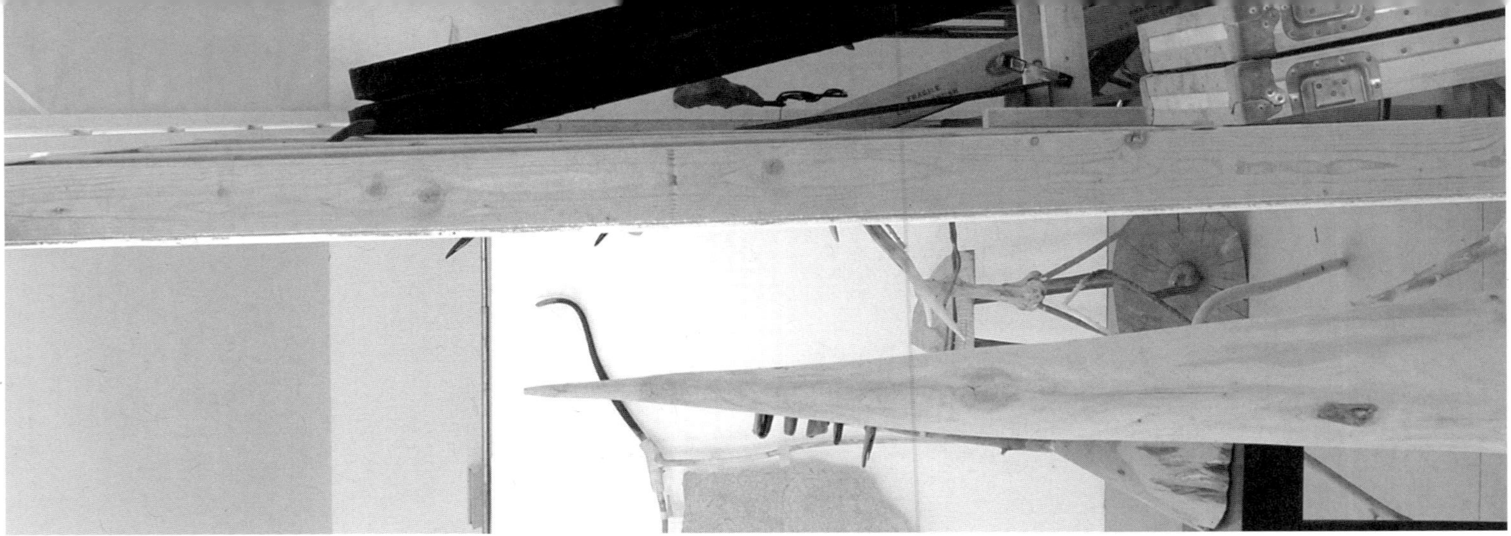

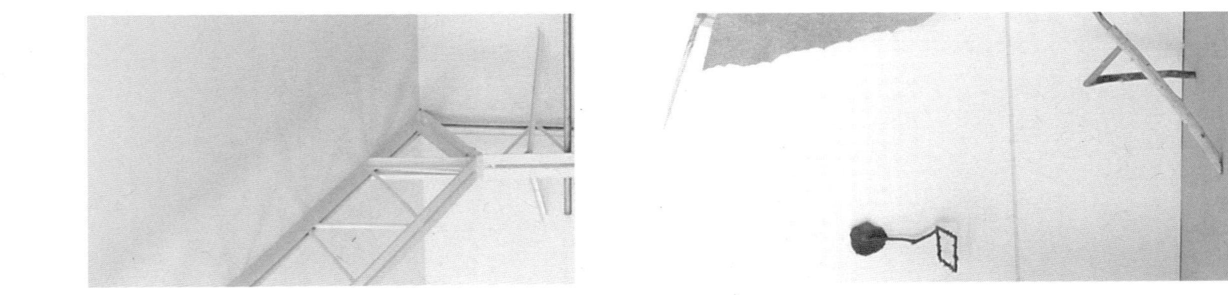

By now the architecture has become white, a backdrop, a gessoed pseudo-blankness, claiming to be absent, against which the activity of the studio wants to be read as a presence. But then, the studio also splits itself into front- and back-of-house, presentable and process, skin and bones. It's no coincidence that walls and paper are both bleached white, of course.

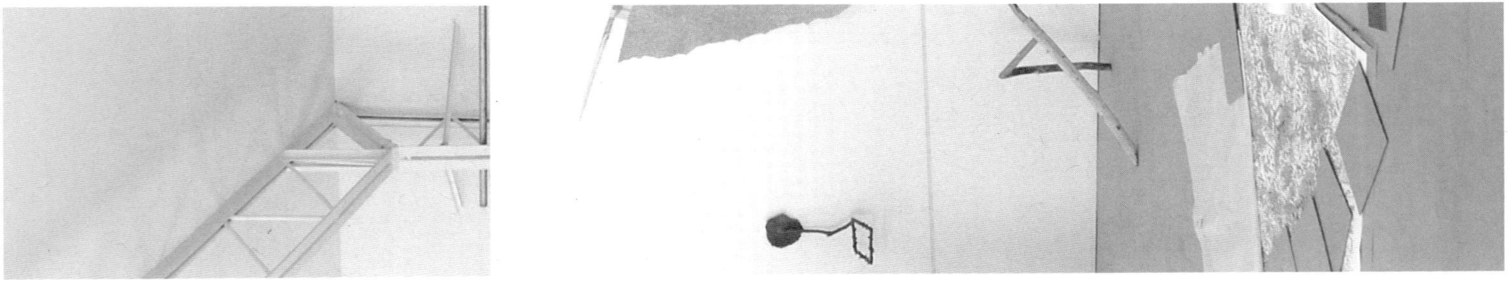

INTERVIEW WITH ARCHITECT
RACHEL ALLEN

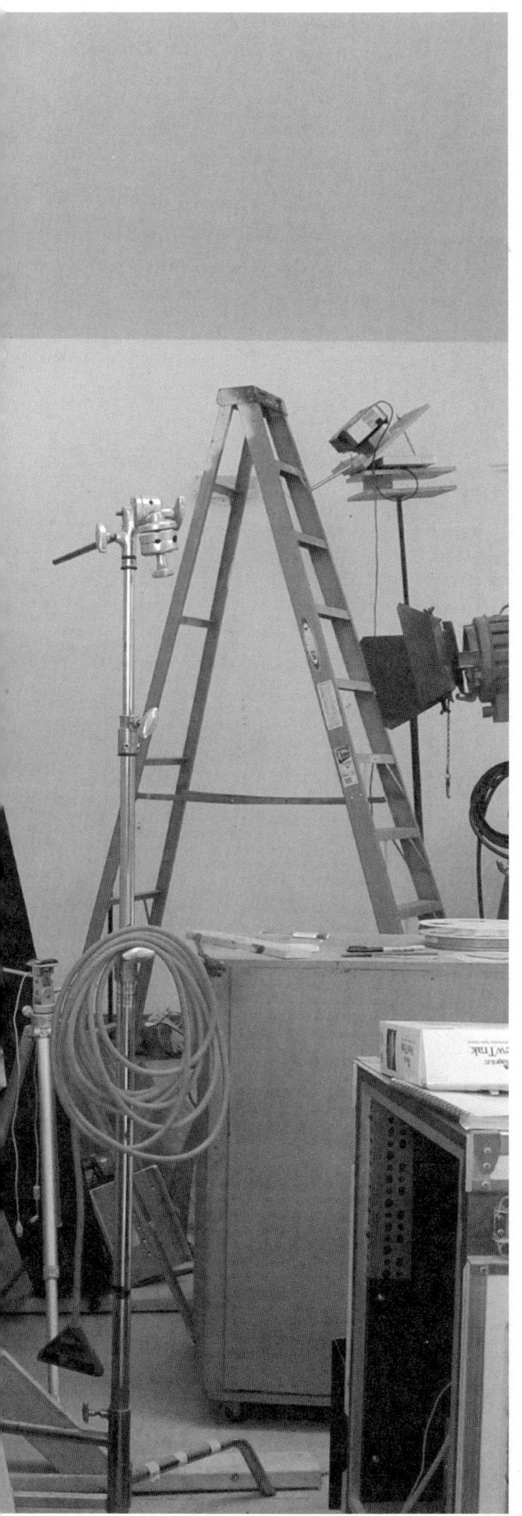

To think about the possibility of interdisciplinarity as it might apply to architectural practices both in the field and in the academy, let me try again. Which disciplines could be intertwined with architecture that haven't been always already, I wonder? Painting, sculpture, and poetry all appear bound up from whatever imaginary origin one might wish to use (murals, monuments, inscriptions). And a building jobsite is so thoroughly collaborative, lately I think of the disciplines I am intertwining with to be masons, drywallers, cabinetmakers.

You write, "a productive model or a useful way to re-think assumptions in a particular discipline." My reply is that it can occasionally be productive or useful, but it can also be lazy or sloppy, as evidenced by your quote from Prix above, and the predominant way architects invoke sculpture. I can also re-think any assumptions I need to from within the parameters and precedents of my own discipline, if I look closely and carefully.

P.S. On "mastery." The profession of architecture is unforgivably retrograde in its attachment to this term. The AIA-LA lecture series is actually still called "Masters of Architecture" and there are so very few women in the profession. But I'm okay with the term in this context, and I know a master when I see one (Ettore Sottsass, Shigeru Ban). It does feel accurate to think of the tools and skills I've acquired over the years as my servants, if not my slaves.

RECEPTION

I should confess my understanding of relational aesthetics is cursory, inherited mostly from overheard dinner party conversations. But that it's a current term, that much I know.

KEN EHRLICH

So I looked up that Benjamin quote again after sending it to you and it turns out I made a critical error in transcription. His statement was not that architecture operates on the unconscious, but that architecture is received in a "distracted state." (Compared to Brecht's idealized state of "distanciation," which is self-imposed.)

Speaking for myself, even at my most self-aware, a portion of myself-as-audience is distracted, and sometimes I find that distracted self to be the most receptive. It is the caught-off-guard self. I might even say it's the audience I am to myself when a good idea comes "out of nowhere" as they say. So I like the idea that my work could continue to parlay that distraction as the circle of the audience widens to include others.

SCULPTURE/WOLF PRIX

I don't believe that architects look to sculpture, because they don't usually look AT sculpture. (I know I didn't, at least not until I met some sculptors.) Rather, I believe they regularly deploy this trope because they believe it is useful to them, in a PR sense. They believe that invoking this figure creates freedoms that they otherwise would not have. The sculptor they are assuming in this invocation is a sculptor who does not and perhaps never did exist—it is a romantic, expressionistic sculptor removed of constraints (constraints which are never removed, and maybe don't even need to be). None of the sculptors I know consider themselves as "free" as the imaginary sculptor invoked here.

Here's another way to say it. If Frank [Gehry] and Wolf are idealizing Richard Serra (the late Serra, softened to fit like a theme park into MoMA and the Guggenheim, not the early one of molten action and the aggression of Tilted Arc), what would it mean if a contemporary architect (say, for example, me) idealized Michael Asher?

The difference between architecture and sculpture? Here's one hypothesis, let's see if I can convince myself of one thing, by writing it out: It is the figure-ground relationship. Architecture (even when it most aspires to be figure) is always background (to action, to function, to occupation). Sculpture (even when it most aspires to be background, as extremely 'relational' as it can get) is still a figure, in the forefront, by virtue of the undistracted attention it expects to receive.

KE: How do you understand the "Los Angeles school" of architecture? How do you relate to the L.A. school in terms of your work? And even more generally, how do you think the city of L.A. has affected your practice?

RA: The LA school of architecture understands itself to be essentially pragmatic. It is understood by others to be essentially anti-intellectual, even willfully stupid.

I understand both interpretations to be flip sides of the same coin. I believe that Frank Gehry learned this stance by observing his artist friends of the 60s—and deducing that the mute or "dumb blonde" artist has always been the critic's favorite kind. If an architect (or artist) talks too much, it just puts the critic out of a job.

I have tried to observe the successes and limitations of this stance and to learn from it. The downside is that whatever success such a stance leads to can always be dismissed as unmerited. Both Frank and Thom Mayne have achieved great stature, but within the architectural community there is a strong prejudice according to which neither of them are taken all that seriously. They are presumed to be some kind of idiot savants. Keep in mind, when an architect gets that "sculptor" tag it is NOT a compliment, it is an epithet.

So, over here at RAA we try not to hide our light under a bushel. We know we're whip-smart, smart-aleky, smarty-pants. And we like to show off, through the ingeniousness of our designs and in our interactions. We've decided to let the chips fall as they may (we do this mostly because we hate being bored.) But being willing to be thought of as a smart architect here in LA is very different than trying to be a smart architect back east. LA gave me permission not to teach, while I built up my practice, for example.

In small and big ways then, the city of LA has made my practice possible, there is no doubt in my mind. There is no other city in which I could convince people to pay for some of the crazy things we want to try, especially not at the age I am and at the stage I am in my career, which is essentially nowhere (that is, the beginning). For this and more, I love Los Angeles with a deep and abiding gratitude.

KE: Beyond the specificity of each project site, can you articulate anything in the way of a methodological approach that you take to considering the qualities of a site?

RA: I have to confess we are shamefully unmethodical. Oh, wait, you said "methodological." Well, I'll start by just telling you the facts. We visit the site. We take a lot of pictures. We build models at several different scales. But we don't really look around much, on the block or in the city. We might look through our library at other architects we like if we remember a solution to a similar problem, or an aspect of the problem. We draw a lot of sections, changes in level and circulation movement, because we're interested in eye contact and its manipulation.

We prefer to consider the qualities of a site in the broadest possible sense. That is, not just topography, plant materials, urban context, regional styles, and so on, but also the precedent of buildings like the one we might like to build, which are located in Russia, Prague, Italy, or St. Louis. I'm reading a lot of Emily Dickinson lately so I might also say that the site with which we are concerned is not primarily "earthly."

Also, we like to imagine that these other buildings will be next to ours in photographs in future textbooks—and that it is through reproduction that buildings are predominantly experienced anyway, not through site visits. We hope to create some kind of conversational continuity between their qualities and ours, and wonder how it can be possible that it feels as if this conversation transcends space and time.

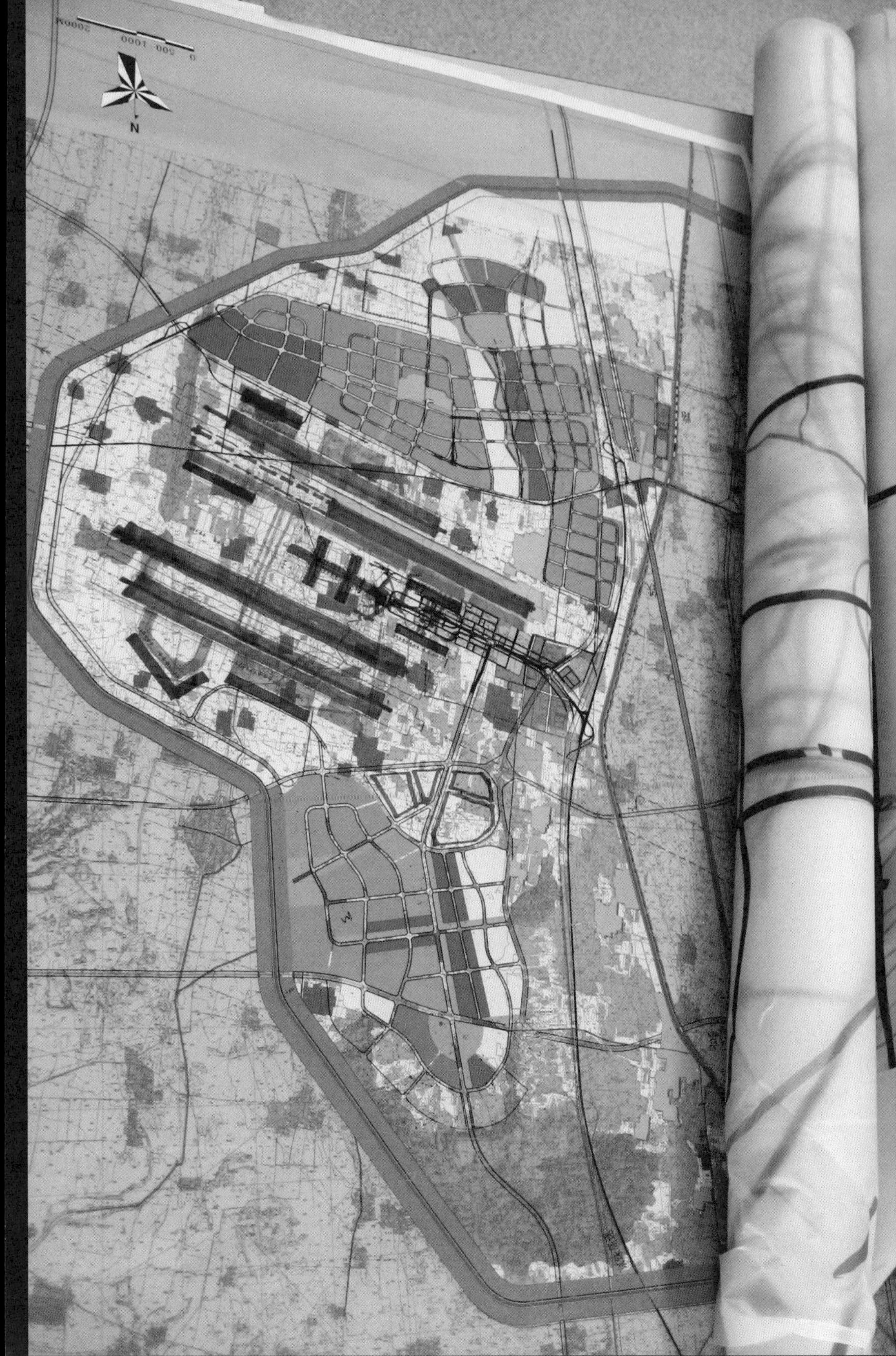

2000m 1000 500 0
N

Ava Bromberg

"What about those brown clusters, aren't they villages? Don't worry about those. Just show us your ideas."
Thoughts on Master Planning in Chinese Cities

MAKE IT AN AIRPORT CITY

The lead planner assembled some junior members of my studio at the China Academy of Urban Planning and Design (CAUPD) around a worktable covered with a large map. The studio was about to start planning an Airport City in Zhengzhou, Henan Province, a prefecture-level city of roughly 7 million people, on the edge of central China's farmland about halfway between Beijing and Shanghai. I had spent the previous week researching international logistics and the world's top 50 cargo airports to provide background for the argument that, given Zhengzhou's location at the intersection of major rail lines and highways, it would be an ideal place for a cargo hub.

One of my colleagues translated as the lead planner placed tracing paper over the map and explained the specifications of the plan area to recent planning school graduates and interns.

"Okay," he said, " Currently there is one runway. Here's where the new river being built to channel water north to Beijing will flow; it makes up the eastern boundary of the plan area. Here's where the major freight rail line runs; it makes up the western boundary of the plan area. Here are the existing major highways. We're building one more major east-west highway. You can put that wherever you want. We need a light rail line. You can put the light rail wherever you want. We need four runways, an industrial area, a commercial center, and a place to house workers. Oh, and we need four clover intersections where the major highways intersect. Okay. Show us your ideas."

I asked, "What about these little brown clusters. Aren't they villages? What about those?"

He replied, "Don't worry about those. Just show us your ideas. You have three days."

There were three weeks to finish the plan for the Zhengzhou Airport City.

Down the auxiliary road sandwiched between the main street and the shops, buses and cars stream along the major boulevard beyond the tree-lined sidewalk to my right. To my left—in front of the shops and between the parked cars—all different configurations of men, women, and children, alone and in pairs, sitting and talking in the dark under a tree or playing on the steps of the empty Bank of China parking lot. People leaking out of their high-rise apartments, claiming small spaces wherever they are available. These inbetween spaces, afterthoughts of rapid development, are socially vital now that all the buildings are finished. The smallest stones and tree stumps are occupied. Multiple generations are walking, talking, sitting together in the sticky summer night. A lone man hunched over his rice bowl shovels food into his mouth at an outdoor dinner table set for four in a parking space. He is backlit by floodlights and the shop-keeper's television across the lot. A few bicycles careen past us, all with little advance warning, in low light made lower by the tree canopy. The street sweepers rake leaves and trash from the sidewalk at 9:30 pm. The soundscape is dense with activities proximal and distant. The smell of grilled meats float up from a temporary sidewalk restaurant. It's late July 2006 on the streets of the oldest "new" Chinese city: Shenzhen, Guangdong Province, People's Republic of China.

Shenzhen exemplifies China's urban explosion. It was a fishing village with less than 30,000 people when it was designated a Special Economic Zone (SEZ) by Deng Xiaopeng in 1979. Today the city boasts an official population of nearly 10 million. Unregistered migrant workers in and around Shenzhen account for an additional 5-7 million.[1] Shenzhen has the distinction of being China's fastest growing city, and its first "modern" city designed mainly around the car, because—as logic would have it—cars are modern. There are no bicycle lanes, and bicycle traffic is relatively sparse compared with Beijing and Shanghai, each said to have six million bicycles. The city is laid out horizontally; super wide boulevards and freeway spines link different ends of the city to the larger city-region. Two subway lines opened in 2004, and a wide network of buses carry most commuters. Within the neighborhoods you can most certainly walk to get around, but cars are ever present and don't yield to pedestrians. It is advisable to look four ways before crossing the street.

Located just over the border from Hong Kong, Shenzhen is home to China's most active mainland seaport. Its clusters of manufacturing facilities make it one of the primary anchors of the Pearl River Delta regional economy, an area often referred to as "the factory of the world." Shoe, plastics, furniture, metal, telecommunication equipment, electronics and other factories abound on the outskirts with intensely dense residential neighborhoods throughout the core and the surrounding areas. Shenzhen is in close competition with neighboring Guangzhou for the title of China's richest city with the highest per capita income; young college graduates flock from their home cities to the center for high-paying jobs, less educated migrants journey from the interior of China to the city's edges for much less well-paid factory and construction work. These

1 Even the City's official website recognizes that the registered / unregistered population structure (with nearly 88 percent of both populations between ages 15 and 59) "polarizes into two opposing extremes: densely populated intellectuals with a high level of education, and migrant workers with poor education." http://english.sz.gov.cn/lis/200509/t488.htm

THOUGHTS ON MASTER PLANNING
IN CHINESE CITIES

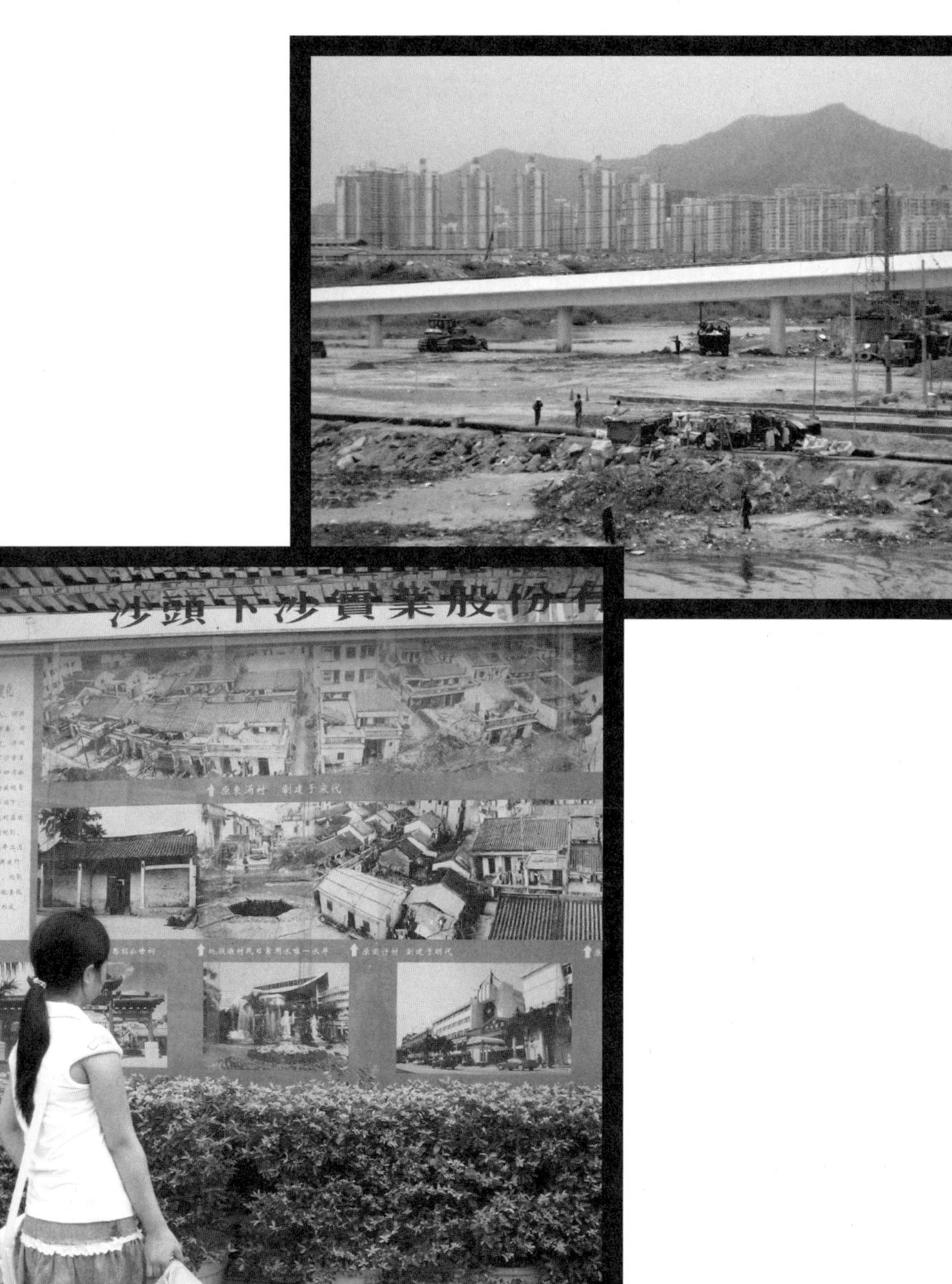

AVA BROMBERG

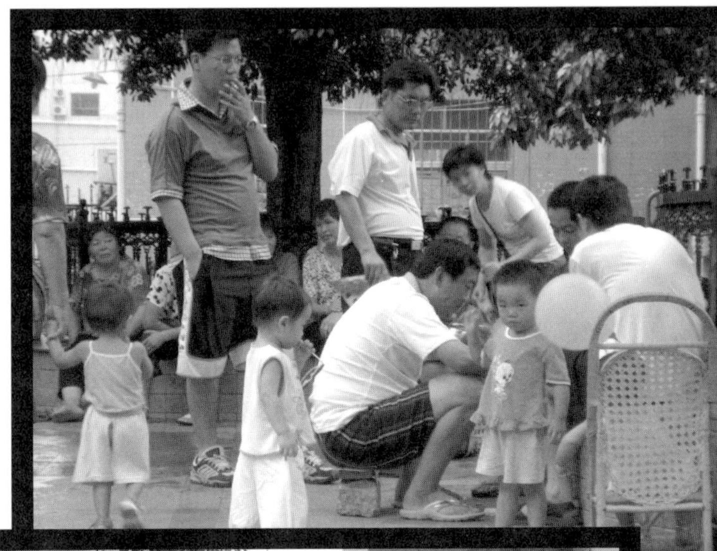

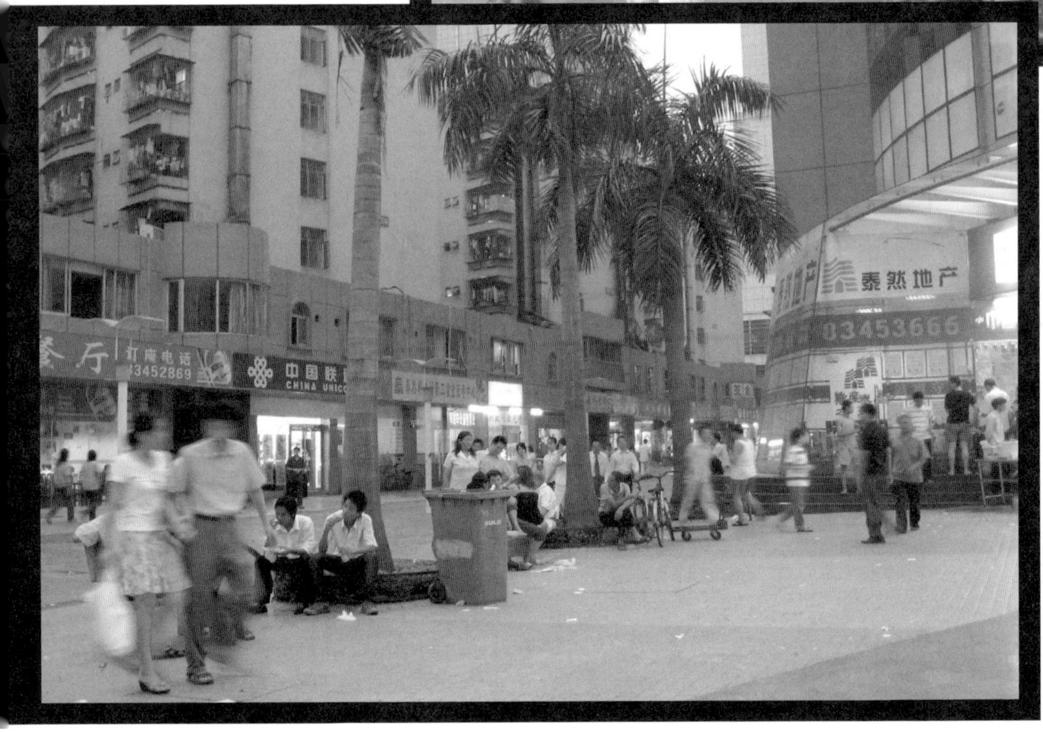

THOUGHTS ON MASTER PLANNING
IN CHINESE CITIES

migrants tend to work long hours to afford their housing and support their families. This bustle is palpable both in the main streets of the city and in the late working hours at the CAUPD planning studios. The daily leisure spaces of the city range from the no-cost and ad hoc repurposing of sidewalks and parking lots to expensive amusement and water parks. Informal uses of open space are plentiful, and seem like treasured activities in such a dense environment. All told, Shenzhen is an overwhelming introduction to the scale, pace, and complications of China's rapid urban development. Especially given that virtually none of the city was here in 1980.

In the summer of 2006, I worked for six weeks in a master planning studio at the Shenzhen branch office CAUPD, the main planning arm of the Ministry of Construction and China's largest planning firm. My job offered an inside perspective on the process of making and remaking Chinese cities from scratch. Daily life outside the studio was rich with information of a different sort. I was left contemplating the contradictions—and opportunities—that this extreme example of city building provides.

My experiences in China highlight a scenario that can be found in many parts of the world, that plays out with unparalleled gravity in China's urban expansion and the rapid pace of planning. It is the tension between the top-down process of master planning and the fact that, at the finer grain of urban life, planned cities will always be transformed through use. While economic and political decisions and priorities take a physical form and set the parameters—the basic canvas of possibilities—after the planning and construction are complete, residents adapt and repurpose the environment. Within the initial constraints people grow their social lives and develop the formal and informal economic activities that turn collections of buildings and streets into neighborhoods. This dynamic may seem obvious enough, but a more subtle point of social and political significance is worth restating—decisions about the built environment shape human possibilities. So in any built environment we might well ask: What capacities remain untapped? Whose desires undeveloped?

In China's case, there are no doubt concrete conditions to improve—millions need housing and jobs, industries need facilities to develop and sell their products, young cities need roads and transit, sewer systems, and park space. Add to this mix mayors competing so their city might push the leading edge of GDP growth. Mayoral decision-making and planning projects they initiate support this primary goal. Established theories of economic geography are thoughtfully considered, but social and environmental concerns less so.

Seeing a major urban area take shape on blank pieces of paper—as if the land itself were blank—is disconcerting. While CAUPD had some sensitivity—at least on the projects I worked on—to preserving the ecology of areas where new factories and dense housing would be built from scratch, there was no engagement with local knowledge or any discussion of public participation in the planning process. And as the Airport City example shows, even where the land is not empty, there are not many insurmountable obstacles. Displaced villagers are simply relocated.

CAUPD has a unique position as a private entity that works in cooperation with the government. The government retains central control over most property in Chinese cities, making master planning a new city (or whole sections of an old one) possible in ways that get very complicated by property rights in the US and other countries. The implications of aggressive master planning, however, are similar in China as they are everywhere. The opportunities to plan more intelligently or blow it completely are great.

AVA BROMBERG

The junior members of the studio worked in multiple directions during the three days we generated ideas for the Airport City. My colleagues leafed through the bookshelf of international architecture and design publications for ideas. For the design of the four-runway configuration they zoomed onto Atlanta's Hatfield Airport on Google Earth, traced the schematic and placed it in the plan area. They lifted ideas and inspiration from German architectural books on the studio shelves as readily as they erased the "little brown clusters." The necessity to come up with ideas quickly trumped any desire to preserve existing forms. The Airport City would create new employment and economic growth opportunities but erase villages without a second thought.

In addition to the Airport City, I worked on the plan for the Shenzhen New Eastern City, a new city for manufacturing on the eastern edge of Shenzhen, to be built on the command of the growth-minded mayor. The old village would be transformed whether the plan was "good" or "bad." At the time, CAUPD was reworking the government planner's environmentally devastating ideas. Instead of redirecting the course of the river, CAUPD suggested leaving it alone, remediating areas polluted by factories and making the river more accessible as an amenity for local residents.

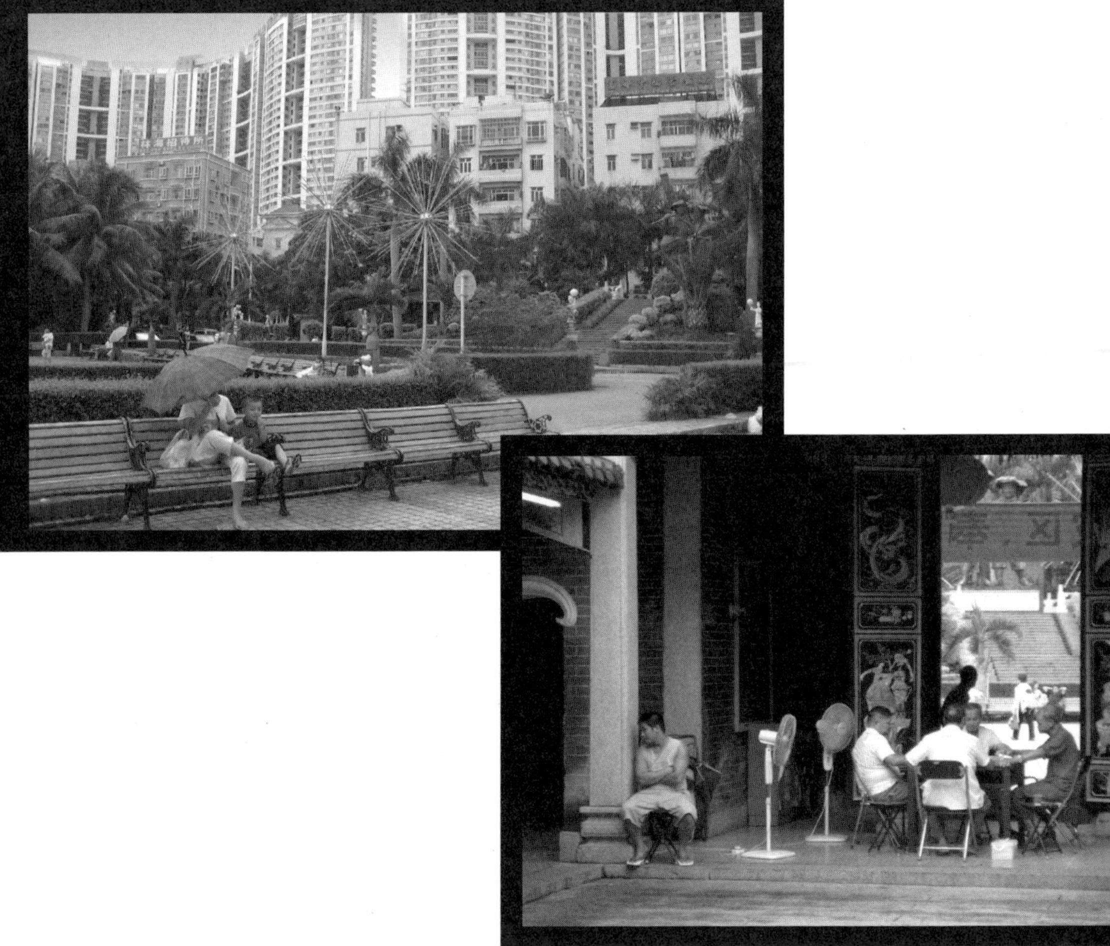

THOUGHTS ON MASTER PLANNING
IN CHINESE CITIES

The aim was to preserve the wetlands as primary filtration for rainwater and floodwater from the mountains. Ideally, CAUPD wanted to set new environmental standards for factories that locate there, to integrate some of the old village structures into new densities and to design the city with alternate modes of transportation in mind (bicycles, pedestrians, buses/trains) instead of cars. The goal was to make it a model industrial cluster town with a better quality of life for the factory workers. Whether this goal is realized, however, largely depends on the mayor, whose vision was to see the area transformed into a "world-class technology park." Indicative of the crazy pace of planning in China, we had twenty days to finish the plan and pitch it to the mayor.

My work on the Shenzhen New Eastern City project began by looking at economic indicators and statistics for the region. For example, by focusing on population growth rate projections, we sought to estimate how many more workers will migrate there from inland areas seeking work in new factories based on previous trends. The growth trends for the years that data was available (1996-2005), were astonishing if incomplete. The kind of statistics the U.S. government keeps—and often makes public on the web—are not openly available in China. Good data can be hard to come by; if they exist they are tightly controlled by the government. It was abundantly clear, however, that unregistered people—meaning those who migrated there from other parts of China for work—made up between 80-95 percent of the population. Locally-born "registered" residents made up the rest. This meant the area had inadequate basic services because schools and hospitals are placed according to the number of registered residents, not the total population. This was a recurring problem in the major manufacturing areas. In planning the Shenzhen New Eastern City, difficulties with the boundaries of administrative regions compounded this problem, so CAUPD's proposal suggested combining three adjacent administrative areas previously treated as separate entities.

After researching economic indicators, I was asked to consider how to connect the public and green spaces between commercial and residential areas with the river that runs through the city. Parts of the river area were to be kept as wetlands—the idea was to have the rest accessible by pedestrians and bicycles. I was asked to create drawings to visualize these linkages. Taking some of the blocky 3-D computer generated images from the master plan as a base, I sketched in people, trees, and some concepts to link these disparate sites. This felt like somewhat of an odd exercise as I remembered throughout the process that these spaces actually come alive once the "planning" is done and people use them. Yet no "people who will use them" were involved in the process. Still, images always help make the sale, and the mayor wanted to see some pictures. With this relatively tiny gesture I kept the scale on the ground human, but was later told that my drawings, while descriptive and well executed, were too detailed.

Three-story structures dwarfed by walls of condominiums behind them. Men playing mahjong in the open-air temple in the middle of the afternoon. Loud clanks and clacks of shuffling tiles. Boisterousness.

A day long tour of Shenzhen began by visiting two of the original villages—called villages-in-the-city (chengzhongcun)—that existed before Shenzhen was designated a Special Economic Zone and sprouted into its current form. Some temples and old narrow market streets remain in the first village, which my officemate and tour guide explained was not just the oldest part of Shenzhen but also the poorest. Though you could not tell by looking, the second village-in-the-city was quite wealthy because of payments received to develop on their land.

AVA BROMBERG

A land of stark contrasts and extremes in built forms: a short time later I'm on a guided tour of a new over-the-top waterfront condo development, soon to open at the rate of 30,100 RMB per square meter. (With the smallest apartment 190 square meters, this translates roughly to $730,000 USD.) This is not just an apartment...the Mangrove West Coast is described as "life well crafted." Replete with super thick energy efficient glass and 30cm thick floors that no bouncing basketball or rambunctious two year old can penetrate, this luxury condo has Jetson-esque conveniences like the ability to control your cctv, the temperature in your house, and your washing machine from the internet. Forget to shut off the coffee pot? No problem. Do it from the office! TV in the fridge, smart phone next to the toilet; what's gained in conveniences seems lacking in warmth. No personality but every apartment is guaranteed that at least two-thirds of the walls have an ocean view. It's with housing complexes like these that Shenzhen hopes to attract more of the business elites and their dollars away from Hong Kong.

The tour finished atop Shenzhen's tallest building, on the 69th floor of the Meridian Tower in the financial district. Shenzhen's density as far as the eye can see (and it was a pretty clear day). Toy-sized cars below in the late afternoon's gathering traffic. Green belts and parks. A stadium in the distance. But just across the river on the Hong Kong "New Territories" side, some rice paddies and farm land, undeveloped mountains, a few clusters of housing towers. In the juxtaposition of dense areas of residential and office towers and stretches of rice paddies, the river in between the two remains a clear illustration of how policy, whatever its inspiration, shaped the landscape over time.

——— ——— ——— ——— ——— ——— ——— ——— ——— ——— ——— ———

The practice of producing a site, of master planning a city from above with the knowledge that inhabitants will transform it and be transformed by it on the ground is an unsettling exercise. Unsettling most of all because it is not an "exercise"—it is a practice with serious consequences for millions of people; it generates the built forms and inbetween spaces that current and future residents will inherit. While in theory this opens up room for wild imaginings about what an ideal city might be, in practice the possibilities are quite constrained. The economic growth that drives—and funds—the process, coupled with the crazy pace and timelines puts serious limitations on what is considered in any plan.

On the streets outside the office I was reminded constantly that, at the core, all the complicated processes—history, government, institutional structures, military might, cultural norms, economic growth, national aspirations—are ultimately important in how they impact people's lives and how people treat each other on a daily basis. I tried to be consistent in the work I did there, the conversations I had, the presentations I made, the perspectives I inserted—that planning (here or anywhere) should try not to forget about people. It's not just that planning should not forget about people, it should think about them. First, last, and always—all the way from the macro-processes to the micro-environment. I harp on it because official planning decisions in China, as in most other countries, come down to economic promise and GDP increasing potential. First, last, and always.

——— ——— ——— ——— ——— ——— ——— ——— ——— ——— ——— ———

On the streets after an evening taking in the Shenzhen club scene, a relatively cool night, the electronics district shops and malls with their shutters down, sidewalks transformed into resting areas for friends to talk or eat street food. My favorite—three older women just sitting in the middle of the sidewalk. No need to move to the side ladies, right there is fine, we'll walk around.

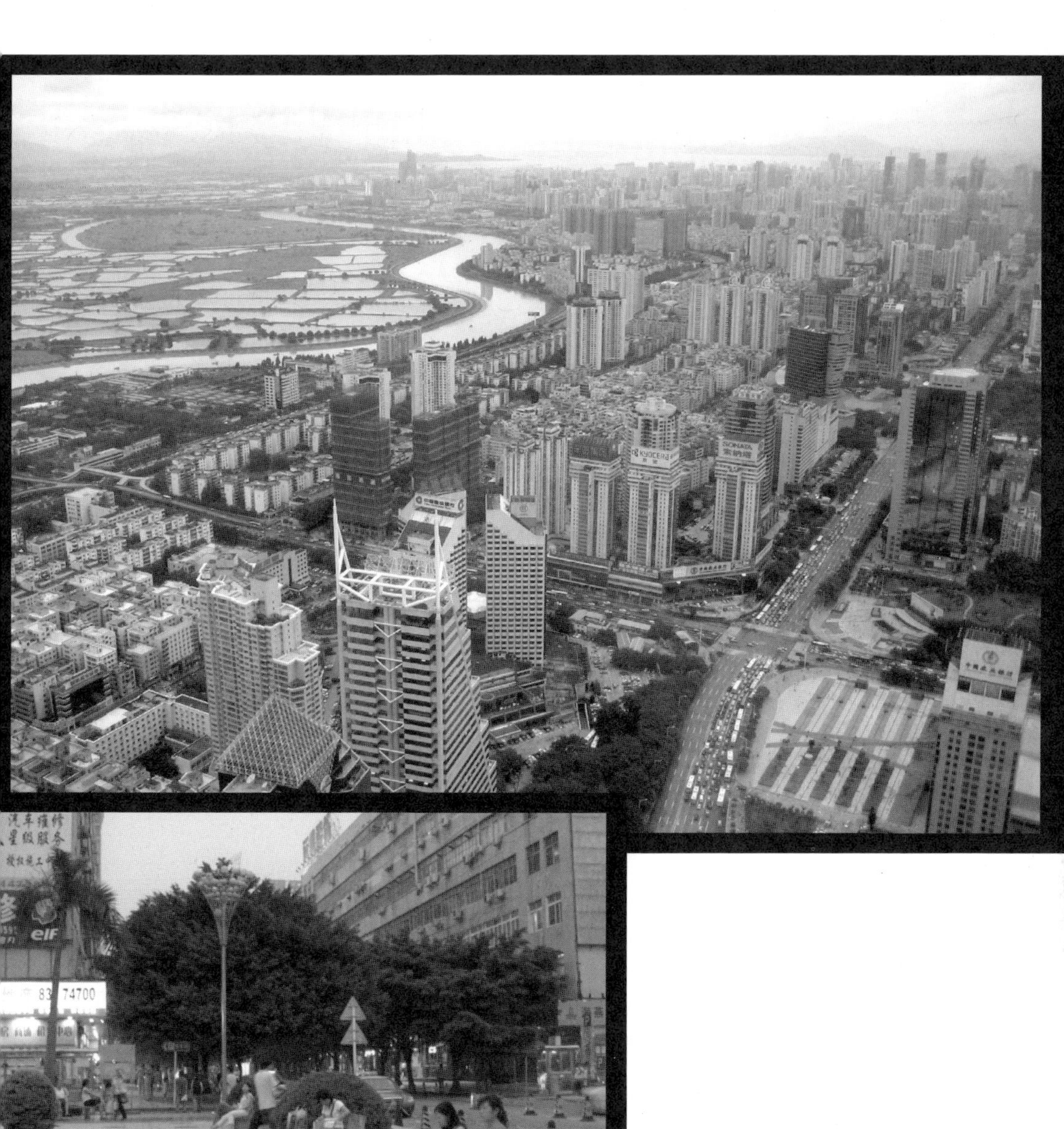

AVA BROMBERG

Seeing them command the sidewalk for seating made my night. Herbal tea stands among the few open storefronts are bright and bustling. The sensory assault of a delicious smell followed by a gross one, from the putrid to the sublime. Late night construction crews still chiseling away, ripping up the floor of a storefront and laying tile under portable florescence.

We remember there are so many big things for "the planners" in China to consider—where to put the roads, the schools, the apartment high rises, the commercial centers. So many things that the details - the tiny details that, in my opinion become the stuff that really counts—get lost or obscured on a consistent basis. Very short timelines and prioritizing growth means very little consideration for the kinds of details that make the microenvironment livable.

My Shenzhen neighborhood was a potent example of how people negotiate and remake spaces that were not designed with their needs and desires in mind. How they make a bad design livable, how they take the structures they inherit or migrate into, transform them and make them their own, how and where they build social opportunities, and where new ones emerge. They succeed in this even where planners gave little consideration to the physical backdrop for this finer texture of social life.

Understanding that new cities are being built every day and few mechanisms exist for protecting the vernacular landscape, questions arise: what elements of the self-generated social environment—the one that's being erased - can be actively transposed into the new forms? How can planners preserve existing social opportunities and create new ones? Can we open up better channels for listening to the people who will inhabit the plan once it is poured in concrete? Can we do so in a way that allows us to retain a critical relationship to our own ideas about how our environments could or should change? What might this look like, and who would dare instigate it in China's current political climate?

The top-down planning that has defined Chinese city-building for a long time, save for the rarely publicized but growing agitations of inhabitants, is neither burdened nor enriched by the kinds of on-the-ground engagement that so often define planning practice in the US and Europe. In Chinese planning, cities are made and villages destroyed by fiat.

Planning is always an expression of power—someone's bottom line—that is given built form. From above or below, with citizen participation or without it, the development calculus likely weighs the needs and desires of those who will ultimately pay the cost of construction with the relative importance of political, economic, social, and environmental intentions that must be heeded to make it happen. It is a shifting, context-specific terrain where key players can steer the course in wildly exciting (but more often predictable) directions.

This dynamic begins to explain the "problem of planning" in general. But there is the other dynamic I introduced at the start: After the "planning" is done, the city built, (re)populated and bustling, the built forms may be simultaneously reified and rendered absurd. They may serve their intended purpose, whether it is to house people, or factories, or leisure activities and also secure a healthy margin of profit. Yet amazingly, even if no one considered what residents might want or need, once people move in they make it their own. The structures may shape how, whether and where people move and relate to their neighbors, but all (the people and the structures) are changed along the way.

THOUGHTS ON MASTER PLANNING
IN CHINESE CITIES

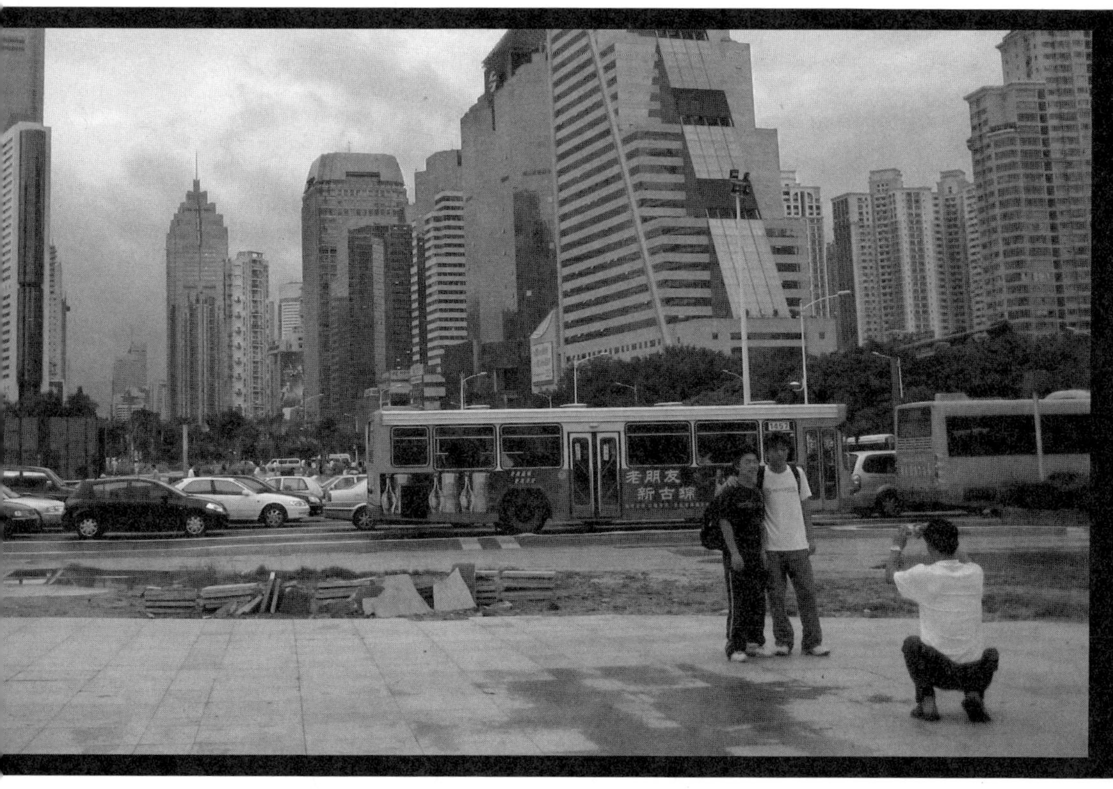

In this way, China remains an extreme case for considering how the initial shaping of an environment shapes the possibilities of future residents. The opportunities and the stakes are enormous, but questions about the long-term social and political impact of planning decisions are still not on most planners' minds. In China as elsewhere, the Master Plan responds to macro-level growth initiatives—the need for more housing, ways to move people, and the country's most recent efforts to advance domestic industry from low-paying production work to high-wage creative work. And in China as elsewhere, the planning process and the spaces it creates are nothing less than the social, economic, environmental and political landscape in which future changes will be instigated.

Last I heard from my colleagues, the Zhengzhou Airport City and The Shenzhen New Eastern City were being constructed, having long since moved from the planners' hands to the developers'. Capacities remain untapped, desires undeveloped, but the finer grain of urban life always has fertile ground to take root.

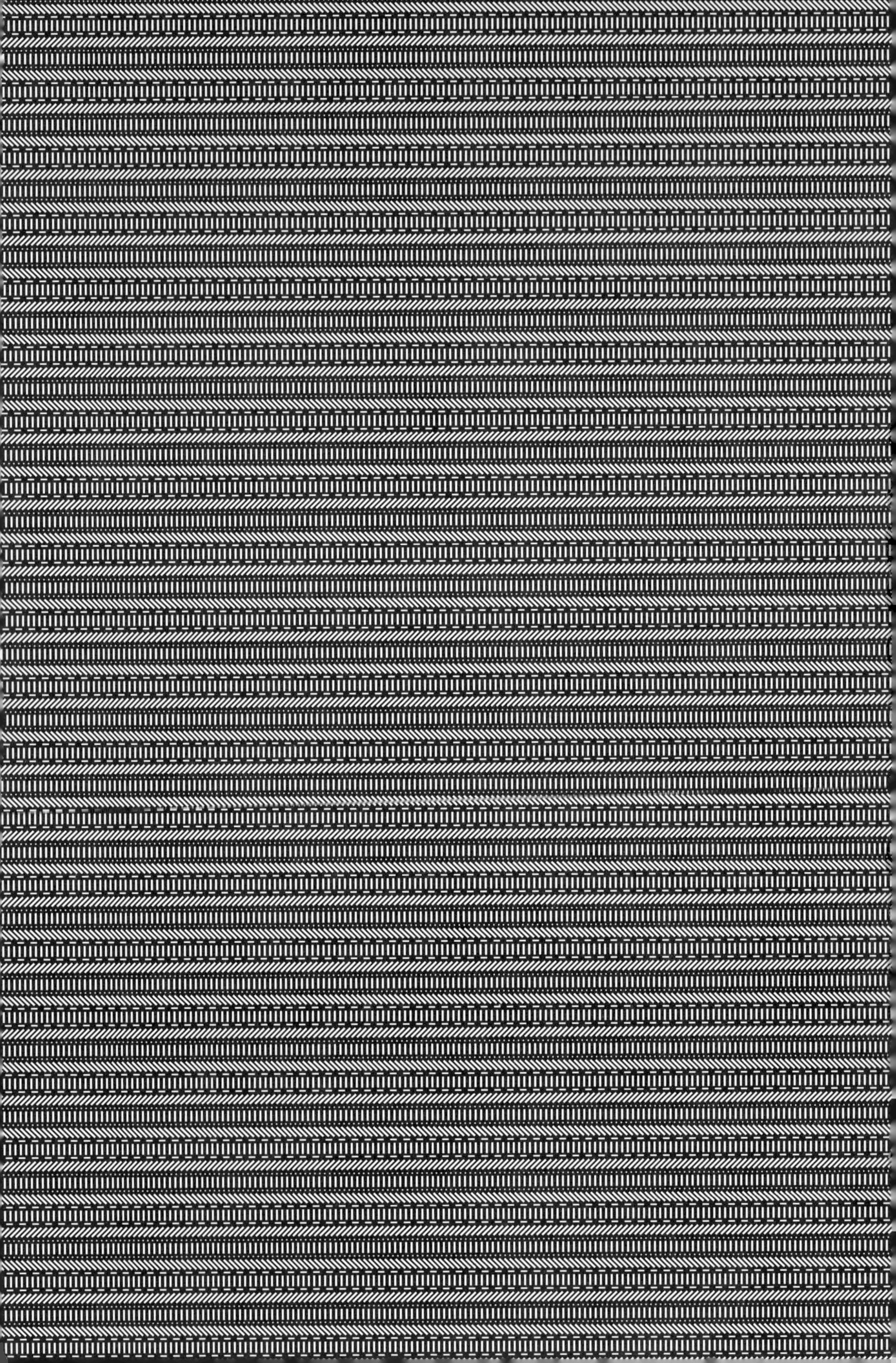